Laboratory Manual for Practical Organic Chemistry

Laboratory Manual for Practical Organic Chemistry

BY

GARFIELD POWELL
Assistant Professor of Chemistry
Columbia University

New York : Morningside Heights
COLUMBIA UNIVERSITY PRESS
1937

COPYRIGHT 1937
COLUMBIA UNIVERSITY PRESS

PUBLISHED 1937

PRINTED IN THE UNITED STATES OF AMERICA

Preface

THE AUTHOR of this manual has attempted to define elsewhere (J. Chem. Educ., 8:107, 568) a method for a first year course in practical organic chemistry. This manual is written primarily for students being taught under that definition: it is hoped, nevertheless, that it will be acceptable to others. We have attempted to give a minimum of specific detail with a maximum of inducement to original experimentation, believing that in the second semester the student is capable of working directly from the literature on some project of his own. We have attempted to emphasize the logical relationship between the properties of the materials involved in a reaction and the manipulations undertaken for the isolation and purification of the desired product. No attempt has been made to illustrate the theory of organic chemistry or any group properties of carbon compounds. It is hoped that the style and content of this manual will be helpful in conveying to the student some mastery of the broad principles of laboratory technique, but even more ardently it is hoped that it will be helpful in interesting the student in a possible application of these principles on some project which will bear comparison truly, even though modestly, to the work of a practising organic chemist.

It is expected that an edition of this manual, enlarged and arranged for general distribution and improved by further collaboration with Professor John M. Nelson, Columbia University, will be ready under joint authorship in the course of a few years.

GARFIELD POWELL

New York, N. Y.
August 1, 1937

Contents

PART I
Exercises

1. Isolation of Caffeine from Tea 7
   (17, 22, 23, 37, 44, 74, 80)
2. Preparation of Ethyl Bromide 19
   (4, 13, 23, 35, 79)
3. Purification of a Known Substance by Crystallization . . . 33
   (17, 22, 36, 56, 80)
4. Preparation of Chloroform 41
   (23, 27, 35, 79)
5. Preparation of Ethylene Dibromide 51
   (9, 23, 34, 35, 47, 79)
6. Preparation of Methyl Phenyl Carbinol 65
   (23, 29, 35, 37, 47, 55, 73)
7. Preparation of Ethyl Benzene 81
   (2, 26, 35, 79)

PART II
Additional Exercises

8. Preparation of Veronal 93
   (17, 23)
9. Preparation of Ethyl Acetate 100
   (6, 23, 35, 79)
10. Preparation of Acetamide 103
    (8, 17, 25)
11. Benzoic Acid from Ethyl Benzene 104
    (17, 36, 56)
12. Preparation of Brombenzene 107
    (17, 26, 35)
13. Preparation of Nitrobenzene 108
    (25, 35, 79)

viii   Contents

14. Preparation of Aniline from Nitrobenzene 109
     (23, 25, 27, 35, 37)
15. Preparation of P-tolunitrile 111
     (13 or 14, 27, 29, 35, 79)
16. Preparation of Hydrocinnamic Acid 113
     (3, 17)
17. Preparation of Acetanthranilic Acid 113
     (14, 41, 42)
18. Preparation of Quinoline 113
     (13, 25, 26, 27, 35, 37)

PART III
Individual Work

Some exercises involving work from the literature or experience
with uncommon reagents and apparatus or careful manipulation
(21, 32, 33, 46, 54, 58, 62, 65, 66, 68, 69, 70, 71, 77, 78, 82) . 117

PART IV
Use of Literature

A List of Books 127
The Library 130
Making a Search of the Literature 132
Beilstein's Handbuch der Organischen Chemie (Fourth Edition) 135

PART V
Qualitative Analysis

Physical Constants 151
Solubility Tests 155
Reaction Tests 156
Tests for the Elements 160
Preparation of a Derivative 161
Purification 162

Index 165

Figures

1. Graded black tea 7
2. (a) Buchner funnel and filter flask; (b) Hirsch funnel and side arm test tube; (c) sintered glass funnel and test tube inside a filter flask 10
3. (a) Squibb separatory funnel for extraction of liquids with cold solvent; (b) Soxhlet extractor for continuous extraction of solids with hot solvents; (c) continuous extraction of liquids by hot solvents of lower density; (d) continuous extraction of liquids by hot solvents of greater density 12
4. Distillation of chloroform from the water bath 14
5. (a) Sublimation on a pad of asbestos; (b) sublimation in a Brühl stand; (c) sublimation in a current of air 15
6. (a) Liebig condenser and air condenser; (b) Liebig condenser made wholly of glass; (c) bulbed condenser for reflux condensation or downward distillation; (d) worm condenser for downward distillation only; (e) condenser with internal cooling; (f) small water condenser; (g) simple reflux arrangement for substances of high boiling point 26
7. (a) Hot water jacket; (b) hot water jacket used as a steam jacket; (c) steam coil jacket 35
8. (a) Simple melting point apparatus; (b) Thiele melting point tube; (c) Fisher melting point apparatus; (d) melting point apparatus with stirrer; (e) copper melting point block 39
9. (a) Steam distillation with a drier; (b) steam distillation with a drier and a Bunsen valve; (c) steam generator 47
10. An assembly of wash bottles: (b) trap for gases; (c) trap for corrosive gases; (d) Pelegot tube; (e) clack valve for air pumps; (f), (g), (h) generators for ethylene 57
11. Addition tube with condenser attachment 69
12. (a) Apparatus for vacuum distillation; (b) small closed tube manometer 73
13. (a) Fractional distillation of crude ethylbenzene; (b) fractional distillation of a substance composed of two individuals boiling at 80°C. and 150°C. 86
14. (a) Hempel column filled with beads or fragments of glass tubing; (b) Widmer column; (c) Young pear column with an attached cooling point for the observation of a reflux ratio; (d) Glinsky column with moving hollow beads 88
15. (a) Air blast stirrer made from a cork and the top of a tin can cut into vanes; (b) mercury seal for a stirrer leading into a closed flask 96

Laboratory Manual for Practical Organic Chemistry

Index of Operations

THE SUBJOINED list is a key to the numbers which follow the title of each exercise. It is sometimes desirable to know what operations a student has completed or how often an operation has been performed or to know what common reagents and pieces of apparatus have been handled. Reference to this list from the numbers on the title of each completed exercise should give the desired information.

1. Alcoholic potash
2. Aluminum chloride
3. Amalgam, sodium
4. Bath, air
5. Bath, metal
6. Bath, oil
7. Boiling point determination
8. Bomb tube
9. Bromine
10. Carbon, test
11. Carbon, determination
12. Chromic acid
13. Cooling, ice
14. Cooling, "dry ice"
15. Cooling, liquid air
16. Condenser, air
17. Crystallization, simple
18. Crystallization, seeding
19. Crystallization, fractional
20. Crystallization, overlaying
21. Crystallization, small quantities
22. Decolorization, charcoal
23. Distillation, simple
24. Distillation, low temperatures
25. Distillation, high temperatures
26. Distillation, fractional
27. Distillation, steam
28. Distillation, superheated steam
29. Distillation, simple vacuum
30. Distillation, about one mm.
31. Distillation, high vacuum
32. Distillation, dry
33. Distillation, small quantities
34. Drying, gases
35. Drying, liquids
36. Drying, solids
37. Extraction, liquids by liquids
38. Extraction, solids by liquids
39. Extraction, continuous, Soxhlet
40. Extraction, continuous, liquids by liquids
41. Filtration, fluted paper
42. Filtration, hot water funnel
43. Filtration, through earths
44. Filtration, through cloth
45. Filtration, press
46. Filtration, small quantities
47. Glass, bending
48. Glass, jointing
49. Grignard reagents
50. Grinding, in mortar
51. Halogens, test
52. Halogens, determination
53. Hydrogen, determination
54. Manometer, closed form
55. Manometer, open form
56. Melting point, determination
57. Molecular weight, by freezing point
58. Molecular weight, using camphor
59. Molecular weight, by boiling point
60. Nitrogen, test
61. Nitrogen, determination
62. Oxidation, electrolytic
63. Phosphorus, test
64. Phosphorus, determination
65. Polarimeter
66. Reduction, electrolytic
67. Reduction, Skita method
68. Reduction, with nickel
69. Resolution, by crystallization

Index of Operations

70. Resolution, with alkaloids
71. Sodium, shot
72. Sodium, wire
73. Sodium, sliced
74. Sublimation
75. Sulphur, test
76. Sulphur, determination
77. Thermometer, correction for b. p.
78. Thermometer, correction for m. p.
79. Washing, liquids by liquids
80. Washing, solids by liquids
81. Washing, gases by liquids
82. Yields, improvement

Part I

Exercises

Exercise 1

Isolation of Caffeine from Tea

(17, 22, 23, 37, 44, 74, 80)

IT IS assumed that the reader is just beginning the study of organic chemistry from the experimental standpoint and that no opportunity has yet been given for acquaintance with the theoretical side of the subject. A formulation of a definite procedure involving chemical properties cannot therefore be profitably undertaken at this stage. Some general methods used by organic chemists can be illustrated, however, leaving the matter of specific reactions to be appreciated after the theory has been mastered.

The extraction of caffeine from tea leaves may be used to illustrate the methods of procedure used by chemists in isolating substances from plant and animal tissues. Caffeine is an alkaloid occurring in tea, coffee, kola, mate (Paraguay tea), cocoa beans and guarana. It is a stimulant and reputedly an antidote for nicotine poisoning. Dried tea leaves of commerce contain about 2 per cent of caffeine, the percentage varying with the grade and class of tea. In preparing black tea for commerce the leaves are graded approximately as shown in the figure.

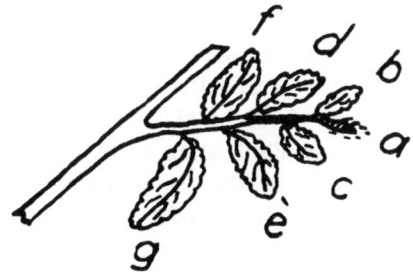

Fig. 1. Graded black tea (about ⅙ natural size).

The leaves are dried, fermented, sunned, "fired" and sieved. The Indian teas (black teas) are usually richer in caffeine but should contain, by virtue of longer fermentation, less of the tannins; the tannins are held to impair or delay digestion to some extent. Adulterants are not common in tea, the principal ones being plumbago (black lead, graphite), foreign leaves, sand, and earths.

A typical rough analysis of the tea of commerce is:

Albuminous matter	25%
Cellulose	20%
Tannins	10%
Mineral matter	6%
Caffeine	2%
Gums, waxes, moisture, etc.	35%

Caffeine is a solid of high melting point and is very soluble in hot water. It is neutral in reaction, giving no salts with dilute mineral acids or with bases. It is remarkably stable, being unchanged by treatment with strong mineral acids—except nitric acid and such oxidants—and even remaining undecomposed when the tea leaves in which it occurs are charred in a flame. It is readily soluble in most organic solvents (2 per cent in cold ethyl alcohol, 0.3 per cent in cold ethyl ether, 13 per cent in cold chloroform), and more soluble in the hot solvents. In hot water its solubility is 50 per cent; in cold water, 1.3 per cent. It crystallizes readily from all solvents.

It is possible to separate caffeine from tea leaves by simply extracting the dry leaves with benzene or chloroform, but, since the purpose of the experiment is to make the operator familiar not with special methods but with methods which are more or less general for the isolation of many neutral substances from plant material, we shall follow the more general method. First comes the preparation of a water extract.

Place the contents of a half-pound package of tea in a very loose bag made of doubled cheese cloth or of wide-meshed linen or cotton and tie the neck. The bag can be made from about half a square yard of such material folded around the tea and tied with cord. Place the bag in a 3,000 cc. beaker containing about 1,500 cc. of boiling water and boil for about thirty minutes, meanwhile heating another 1,000 cc. of water for a second extraction later. Press the bag occasionally with a glass rod, and, at the end of the given time, transfer the extract to a large dish or beaker and squeeze the bag well in order to leave as little water as possible adhering to the leaves. When heating the water in the large beaker use a small flame at the beginning: large glass utensils are usually thick for the sake of mechanical strength, and thick glass, even though made of Pyrex, is liable to crack with large temperature differences at the faces. It will be seen that the volume of water extract is much less than 1,500 cc.: for this reason we extract once more with the second 1,000 cc. of water, again boiling for thirty minutes and squeezing the second ex-

Isolation of Caffeine from Tea

tract also into the dish. It will be agreed that, for such a large surface of material as is presented in the swollen tea leaves, the volume of water here employed is none too large for efficient extraction.

The water extract will contain, in addition to the caffeine, much albuminous material, tannins, and some mineral matter. The other materials designated in the table are insoluble in water. The tannins and albumens, always present in such plant extracts, give insoluble precipitates when treated with lead acetate or litharge: since caffeine does not act in this way, precipitation and filtration afford a means for eliminating tannins and albumens from the solution.

While the second extraction is proceeding, make up a solution of basic lead acetate. Dissolve 80 g. of lead acetate in 500 g. of water. Bring to the boil and add, in small portions with stirring, 80 g. of litharge (lead monoxide). Boil for a few minutes longer and filter while warm through an ordinary filter funnel. Not more than 200 to 300 cc. of clear filtrate need be collected, the residual solution and also the residual insoluble lead compounds being used in the first part of the precipitation of the tannins, etc., and the clear filtrate being reserved for the end of the precipitation. This method of using the lead compounds is suggested simply because the use of litharge alone involves a difficulty of seeing the end point, litharge being insoluble and hard to distinguish from the precipitated tannins and albumens.

Transfer the collected tea extracts into a 3,000 cc. beaker, rejecting the last 20 or 30 cc. if a muddy sediment has collected at the bottom of the dish. We shall lose only a very small proportion of water solution of caffeine in so doing and it matters little what the sediment is since we know the caffeine is in the water solution: the chemist takes every opportunity for removing unwanted material. Keeping the extract hot, but not boiling, add the residual insoluble lead compounds and residual solution of basic lead acetate. Stir with a rod. Let the precipitate settle. Test the clear supernatant liquid with a drop of the clear solution of basic lead acetate; this can be done by allowing a drop or two of the lead acetate solution to run down a glass rod touching the side of the beaker just above the surface of the liquid. If a precipitate forms, add 10 cc. to 15 cc. of the solution of basic lead acetate. Stir again, and, keeping the extract hot, again let the precipitate settle. Test as before and continue until a test gives no precipitate or only a negligible one. This end point will be rough. Remove the flame. While the beaker is still hot to the hand (60°C. is about the maximum that can be borne without discomfort on long contact), filter off the precipitate on a large Buchner funnel connected to the pump and press out the mass with a spatula or glass stopper.

A 4" Buchner funnel may be used and the material may be divided into two portions. The pressing should be thorough, since the voluminous precipitate absorbs a large quantity of water.

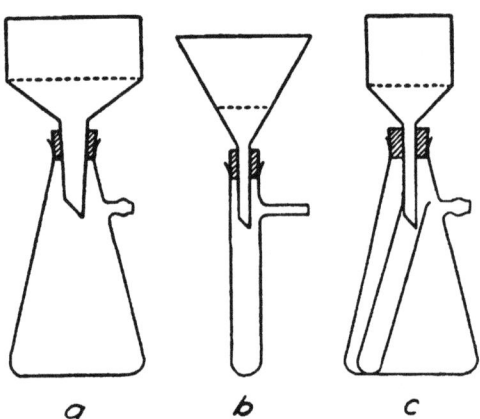

Fig. 2. (a) Buchner funnel and filter flask; (b) Hirsch funnel and side arm test tube; (c) sintered glass funnel and test tube inside a filter flask.

The filter paper should be seated flat on the funnel and should be moistened before the filtration is begun. This moistening, followed by suction, makes for a close fitting of the paper to the surface of the Buchner and helps to prevent particles creeping around the edges of the paper into the first portions of filtrate. Often the first portions of filtrate have to be refiltered in any case, the bed of material being thereafter sufficiently dense to make the filtration satisfactory: the bed, indeed, is considered to be the filtering agent of major importance. In dealing with solvents which do not soften the paper it is sometimes advantageous to moisten the paper first with water and then to wash it with a solvent which mixes with the given solvent and with water, and finally to wash it with the given solvent: in this way the paper holds the first close conformation to the surface of the funnel that the moistening gives it (filtration, pp. **8, 34, 35, 105**).

The residue on the funnel is now washed twice with about 200 cc. of hot water (200 cc. merely because the large bulk of precipitate impels the use of large quantities of wash material).

The washing is done by first disconnecting the pump (the pump is not turned off because this might lead to a backing of water into the filter flask) and then adding the hot water to the material on the funnel. The mass is

Isolation of Caffeine from Tea 11

stirred with a rod till uniformly wetted and the pump again connected, the mass being again pressed well. Repeat with the next 200 cc. of hot water. This washing is not thorough. For such compact material it would usually be best to remove the material into a beaker and add the wash water, stirring well, warming perhaps, and refiltering, and moreover employing three or four portions of wash water, not one or two. It may be well to note here that a given volume of wash material is most efficiently employed when divided into as many portions as possible to be used in succession (washing, pp. 10, 36, 59, 63).

The filtrate contains the caffeine. Bring the filtrate once more to boiling and add dilute sulphuric acid (10 per cent or less by weight) until no further precipitate of lead sulphate is formed, proceeding as before except that, towards the end, the sulphuric acid is added drop by drop. With the solution still hot, decant away from the lead sulphate or filter away. If it is decided to filter, use the Buchner funnel in the manner described in the exercise on crystallization (Exercise 3). This is the second occasion in which the instructions have called for a hot precipitation.

It is nearly always the case that the precipitate formed by a hot precipitation is easier to filter than that obtained without heating: some process of coagulation of small particles, or removal of traces of volatile solvent, etc., leads to this result; rarely is a precipitate less soluble hot than cold. The hot filtration naturally follows the hot precipitation when the substance precipitated is less soluble hot than cold, or when there is danger of some reversal of the coagulation, or even when it merely saves time (filtration, pp. 8, 34, 35, 105). Hot filtrations are most often called for in crystallizing, as illustrated in Exercise 3.

The lead has now been removed and we are left with a very large volume of solution containing a small amount of caffeine and soluble inorganic impurities. For the sake of convenient handling in the later extractions and crystallizations it would seem proper to concentrate, if possible, at this stage. The simplest way to concentrate would be to evaporate the water over a flame in a large dish or dishes wherein the large surface of water is helpful in expediting the evaporation. This will be allowable only if caffeine is not appreciably volatile in steam, as it happens not to be. Concentrate to about 250 cc. Cool. We can now separate the caffeine from the water and inorganic impurities by the use of an organic solvent which is not miscible with water but which does dissolve the caffeine. Chloroform and ether are such solvents, and, by reference to the solubilities before mentioned, it can be seen that chloroform would be the better for our purpose: this because the ratio of the concentrations of caffeine in water and the extractant liquid will be the ratio of

Isolation of Caffeine from Tea

its solubilities therein and caffeine is far more soluble in chloroform than in ether. Extract with 50 cc. of chloroform in a separatory funnel, as shown (solvents, pp. **33, 40, 105**).

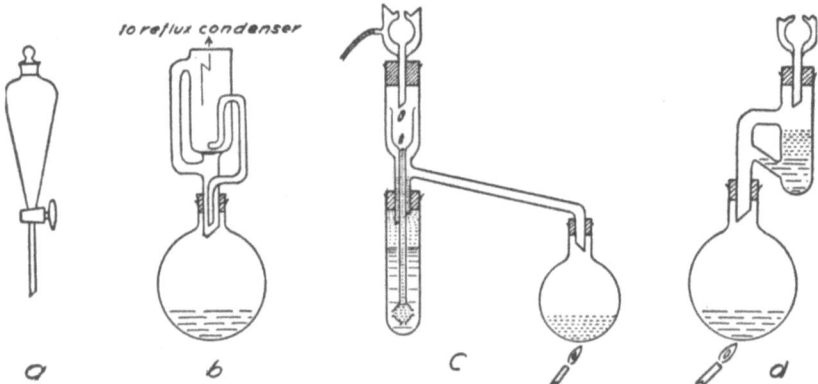

Fig. 3. (a) Squibb separatory funnel for extraction of liquids with cold solvent; (b) Soxhlet extractor for continuous extraction of solids with hot solvents; (c) continuous extraction of liquids by hot solvents of lower density; (d) continuous extraction of liquids by hot solvents of greater density.

Do not work with the funnel more than two-thirds full in the process of extraction (if necessary divide the water solution into two portions). Close the funnel and invert it. Release the pressure by opening the stopcock. Shake gently. Again open the stopcock. Repeat until there is no sound of escaping vapor on opening the stopcock. It is now safe to close the stopcock permanently and shake the funnel vigorously up and down to mix the liquids thoroughly. Shake for a minute or so, the more vigorous the shaking the more satisfactory being the extraction. Then open the stopcock and run off the lower layer of chloroform into a convenient receptacle. Repeat with two more portions each of 50 cc. of chloroform and collect the extracts together.

In the use of extractant, as in washing, the greatest efficiency is obtained by using a given amount of extractant in as many fractions as possible. The amount of extractant and the number of times it should be used can be decided by comparing the solubility of the material in the original solvent with its solubility in the extractant. Assuming immiscibility of extractant and solvent, and a solubility of the material to be A per cent in the extractant and B per cent in the original solution, then the ratio of substance in extractant to substance remaining in the original solvent will be A to B, if equal volumes are taken. If x grammes of extractant are used and y grammes of origi-

Isolation of Caffeine from Tea

nal solution, then the ratio will be Ax to By. The fraction extracted is therefore $\frac{Ax}{Ax + By}$ Using the same amount of extractant in each succeeding extraction, we shall remove the same fraction of the remainder in each case and it can easily be proved that, if F is the fraction and the extraction is done N times, the total fraction extracted will be $1-(1-F)^N$. In this case the solubility is 13 per cent in chloroform, 1.3 per cent in water and we use 75 grammes of chloroform and 250 grammes of water. The ratio in the layers is therefore $\frac{13 \times 75}{1.3 \times 250}$ which is 3 to 1. The fraction extracted is therefore 3/4. The next extraction will take out 3/4 of the remaining 1/4, leaving 1/16. The next extraction will take out 3/4 of 1/16, leaving 1/64 only of the original material. The formula, which is hardly worth memorizing, gives the same result. In practice, the answer is only approximate for most cases and the process is sometimes complicated by the factor of compound formation between the material and the original solvent or extractant (extraction, pp. **29, 30, 40**).

We have at hand a solution of approximately 5 g. of caffeine, fairly pure, in about 175 cc. of chloroform, a solvent which boils at 61°C. The boiling point of caffeine is about 250°C. and we can presume that it will be safe to drive off the chloroform without loss of caffeine. A temperature difference of 100°C. between the boiling points of two organic compounds is, as a rule, difference enough to justify the assumption that only a small quantity of the higher boiling substance will be distilled at or near the boiling point of the first material (distillation, pp. **14, 24, 70, 72, 82**). The chloroform can be distilled, without waste, on the steam bath and collected in a receiver as shown in the diagram, the corks being softened as described later (p. **25**) and the clamps being only just tight enough to hold the apparatus in position.

Never use very rigid assemblages of apparatus: glass is very brittle.

Before beginning the distillation it is advisable to place in the distilling flask a small piece or pieces of porous tile or hard coal to prevent bumping: the pieces need not be much larger than a pin head. This is a precaution against superheating and consequent bumping when the vaporization does take place, the bumping being due to a sudden and extensive vaporization which is violent enough to disturb the flask and, often, to force liquid and vapor in a strong stream through the exit tube. As most organic liquids are inflammable, there is danger of fire as well as waste of time or material in this event. The tile is effective because its pores maintain air at the place where superheating is taking place—the bottom of the flask—and it therefore induces the formation of the vapor phase of the material being distilled (distillation, pp. **74, 82, 153**).

In Exercise 6 will be found an even more effective device for inducing boiling. The distilling flask should be of the high side-arm type rather than another, though it is not an important matter here. When dealing with liquids of high-boiling point (about 300°C.) the low side-arm type of flask is usually chosen.

In every distillation a certain temperature difference must be maintained between the source of heat and the vapor distilling over, a difference sufficient to overcome the heat losses at the surface of the flask. This "head" of temperature is usually about 20°C. to 30°C. for medium flasks at temperatures around 100°C. or 150°C., meaning that the source of heat must be about 20°C. to 30°C. higher than the distilling temperature. With larger flasks it may well be 40°C. to 50°C. At higher temperatures, however, in which the heat losses increase in more than simple proportion, the "head" of temperature may have to be 70°C. or 100°C. for medium flasks. This may well involve danger of decomposition of the organic material because the material at the bottom of the distilling flask approaches in temperature the temperature of the source of the heat itself. Hence the employment of a low side-arm flask, the condensing surface being much smaller and the necessary temperature difference being also smaller. The less the cooling surface, however, the less perfect the separation of the constituents of the mixture being distilled. (distillation, pp. **82, 84, 107**).

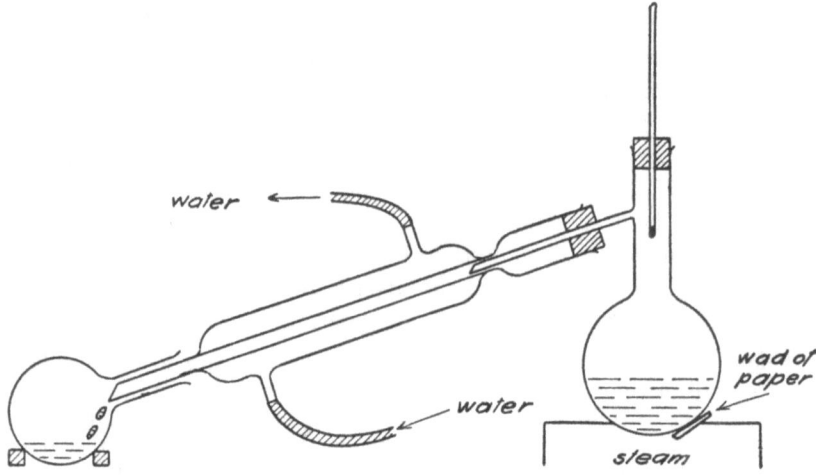

Fig. 4. Distillation of chloroform from the water bath.

The diagram shows a thermometer in the correct position for a distillation—the top of the thermometer bulb on a level with the bottom of the side-arm opening. The source of heat is 100°C. (steam) and chloroform boils at 61°C., ample head for driving over the chloroform with no danger of driving over the caffeine, the end of the distillation being shown when no more chloroform condenses in the receiver. When the

Isolation of Caffeine from Tea

chloroform has been driven off, the distilling flask containing the solid residue of impure caffeine is detached (cork boring, see p. 25).

The last step in purification is the sublimation of this caffeine, a procedure in which a small crucible is used as a container for the caffeine. We must transfer the caffeine to the crucible. It is obvious that removal of the caffeine as a solid would be difficult; a more convenient way will be to dissolve the caffeine once more in a little of some solvent. While doing so, it will be just as well to crystallize and so add another purification step. Add 15 cc. or 20 cc. of water to the caffeine. Warm to dissolve. Transfer to a beaker or dish of 100 cc. or 150 cc. capacity. Wash the flask twice with 15 cc. to 20 cc. of water, pouring the washings into the beaker and so swirling the wash water and draining into the beaker that the minimum amount of caffeine is left behind in each transfer. Now evaporate the solution with a small flame under a gauze, stirring to prevent bumping or spitting, until the volume is about 15 cc. Cool. Since only about 0.2 g. of caffeine is soluble in 15 cc. of cold water, the rest will crystallize from solution. Remove the adhering water (or rather, caffeine solution) by filtration and by pressing in a small Buchner funnel, or merely by pressing on a large sheet of hard filter paper or ordinary note paper. Then transfer the caffeine into the crucible and press it down.

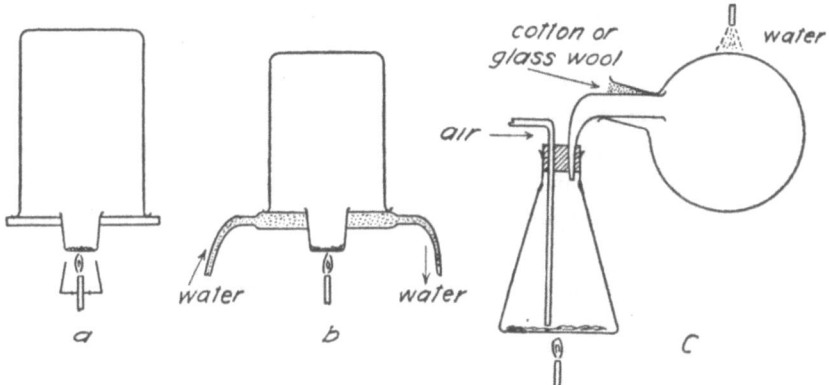

Fig. 5. (a) Sublimation on a pad of asbestos; (b) sublimation in a Brühl stand; (c) sublimation in a current of air.

The crucible is sunk well down in the hole in the center of the asbestos pad and the heating is done with a small blue flame protected by a chimney. A filter paper, cut to fit around the hole in the pad, rests on the pad and is surmounted by a large beaker (700 cc. to 1,000 cc.). If the flame

16   **Isolation of Caffeine from Tea**

is small and the crucible well down in the pad, the caffeine which sublimes near the crucible will not be re-melted by the heat. The crystals from the water solution being hydrated, water condenses at first; it should be removed with a towel. Avoid breathing the fumes. Finally the caffeine sublimes and only a very small black smear is left in the crucible. The caffeine so prepared is almost pure. Caffeine melts at 234°C. to 235°C. and boils a few degrees higher. The yield varies with the grade and kind of tea and the efficacy of the operations of extractions, washing, etc. The average yield is 2 g.

Sublimation is a kind of distillation. Most substances melt before they distill at atmospheric pressure but some substances have a melting point either above the boiling point or very little below it. Such substances, when distilled from a flask, clog the exit tube badly and make some such practice as the foregoing a necessity if purification is to be attempted through vaporization and condensation. It will be noted, later, that the practice is not in general as convenient or dependable as distillation.

THE LABORATORY NOTE BOOK

Each student should be provided with a firmly bound laboratory note book. Every exercise should be understood before proceeding to work and the note book should be prepared for entry with equations, tables, headings in which the operations can be noted immediately as performed, and an abstract of the instructions written out so that actual operations can be compared with them immediately when done. Mistakes, yields, and references to the literature consulted should also be noted for the information of the instructor. Do not hesitate to ask questions. Abstracts of the original literature, when used, need not be kept in the laboratory note book but they should be available to the instructor at all times. References to the original literature should, however, appear at the end of the notes.

No student will be allowed to proceed until the note book is prepared for entry and the laboratory manual is put away.

The following example illustrates the preparation of the note book for the exercise just completed.

Exercise I. Isolation of Caffeine from Tea

Method. Extract the caffeine, tannins, albuminous material, inorganic water soluble impurities with hot water. Remove the tannins and albumens with lead acetate as insoluble lead "salts." Extract the caffeine from the water solution with chloroform and so eliminate water-soluble

Isolation of Caffeine from Tea 17

inorganic material which is usually insoluble in organic solvents. Remove chloroform by distillation. Sublime the residual caffeine.

Detail. Extract ½ lb. tea. (Jones "Delecta") in a loose cloth bag with 1,500 cc. water by boiling ½ hour. Then squeeze and extract again, similarly, with 1,000 cc. water. Squeeze again.

$$\text{Used} \ldots\ldots \text{cc.} + \ldots\ldots \text{cc.} + \ldots\ldots \text{cc.}$$
$$\text{Boiled} \ldots\ldots \text{hr.} + \ldots\ldots \text{hr.} + \ldots\ldots \text{hr.}$$

Precipitate tannins, etc. 80 g. lead acetate in 500 cc. hot water. Stir in, boiling, 80 g. litharge. Filter about half of the solution. Keeping the extract hot in a beaker add, in portions with stirring, the insoluble lead compounds and residual unfiltered solution. Finish precipitation with the clear solution.

$$\text{Unused solution} \ldots\ldots \text{cc.}$$

Remove the precipitate on Buchner funnel in two parts. Wash each part with 100 cc. and 100 cc. of hot water. Press well between washings.

$$\text{Used}\ldots\ldots\ldots$$
$$\text{Pressed}\ldots\ldots\ldots$$

Bring filtrate to boiling. Precipitate excess lead as sulphate with 10 per cent sulphuric acid. Avoid excess.

Actually 11 per cent sulphuric acid.

Excess 3 drops?

Used ..... cc. of the sulphuric acid.

Decant or filter hot. Decanted.

Evaporate to 250 cc. Actually to ..... cc. Extract with 50 cc. + 50 cc. + 50 cc. chloroform in succession. Actually used: ......

Drive off chloroform on the water bath. Porous tile. 500 cc. flask.

Crude caffeine taken out with 15 cc. + 15 cc. + 15 cc. water in succession.

Cool to crystallize, after evaporation to 15 cc., and filter. Press.

Place in crucible in apparatus as shown in the manual. Sublime with a small flame. Remove water as it comes off. Avoid breathing the fumes.

Weigh in bottle. Yield 3.4 g.

Notes. Yield is 20 per cent greater than average. The extraction or the washing more times than given in the manual the cause? Which? Both? The residue in the crucible about average. When extracting with chloroform the water was not cooled to room temperatures and the first shaking led to loss of a small amount of chloroform when the stopcock was opened. Chloroform b.p. about 60°C. Note this with other volatile solvents: ask the instructor about the common ones, ether, alcohol, benzene, acetone, etc.

Why avoid excess sulphuric acid? Is it extracted with chloroform? If not, why avoid excess? Most inorganic salts and bases except ammonia are only very slightly soluble in organic solvents. Acids perhaps more soluble?

If excess sulphuric acid finds its way in part into the chloroform solution it will be left behind on the caffeine. Most organic compounds on heating with conc. H_2SO_4 are decomposed.

Why not shake the chloroform solution with sodium carbonate solution once to neutralize any sulphuric acid in chloroform solution? (This would be done, usually, with substances not so extraordinarily stable as caffeine.)

No literature except the manual.

Exercise 2

Preparation of Ethyl Bromide
(4, 13, 23, 35, 79)

IN THE course of work the organic chemist often has occasion to convert one compound into another. Very often the conversion is a matter of replacement of one atom or group in the molecule of the original substance by another atom or group, the new atom or group being a part of some reagent which is made to react with the original compound. One of the common transformations of this sort (double decomposition or metathesis, or mutual exchange of atoms or groups) is that of an alcohol into an alkyl halide, the hydroxyl group of the alcohol being replaced by a halogen atom derived from a halogen acid: $ROH + HX \rightarrow RX + H_2O$ (X being halogen). This formal representation of the reaction is very simple but the detail of the procedure undertaken in the laboratory may be quite complicated, depending on the properties of the particular alcohol and halide under consideration. The textbooks of organic chemistry mention other reagents which can be used for the transformation of an alcohol into an alkyl halide: the choice of the reagents, as well as the detail of procedure, will depend on the properties of the particular alcohol and halide.

Reagents and Procedures

The choice of the reagent may depend on such factors as the class of the alcohol, the molecular weight of the alcohol, and so on. If it were desired, for instance, that a certain alcohol be converted into the alkyl chloride, it would not be sufficient merely to know that hydrochloric acid, phosphorus pentachloride, or phosphorus trichloride is the reagent generally recommended for the conversion of alcohols to alkyl chlorides. It has been observed that hydrochloric acid gives the chloride from tertiary amyl alcohol and tertiary butyl alcohol with remarkable ease at low temperatures, whereas it gives the chloride of n-butyl alcohol and most of the other amyl alcohols only with difficulty at higher temperatures: the class of the alcohol (primary, secondary, or tertiary) is apparently worth thinking about. The molecular weight of the alcohol may be important: it is a fact that phosphorus trichloride gives little or no yield of chloride with methyl alcohol and some other simple alcohols.

The details of the procedure depend largely on the physical properties of the reagents and products, particularly on such properties as boiling points, melting points, and solubility. The case of ethyl alcohol and the ethyl halides illustrates the matter for the simpler alcohols. Ethyl chloride is a gas at room temperatures, but ethyl bromide and ethyl iodide are liquids: the apparatus used in the preparation of the chloride would of necessity be different from that used in the preparation of the bromide or iodide. For ethyl bromide we could use hydrobromic acid but in the case of the iodide it would be more expedient to use phosphorus tri-iodide: this because hydrobromic acid is fairly stable in solution and could be obtained from stock or commercial sources whereas hydriodic acid is not stable and could hardly be obtained pure from the usual sources. The phosphorus tri-iodide could be generated by the use of phosphorus and iodine in the reaction flask with ethyl alcohol. In the case of the bromide, indeed, we could even avoid the use of hydrobromic acid solutions: we could use a mixture of sodium or potassium bromide, sulphuric acid, and alcohol, materials found in all laboratories. The reaction between sulphuric acid and the sodium or potassium bromide would furnish the hydrogen bromide in situ. Such a method, however, would not be practical for the preparation of ethyl iodide since, as we recall, the hydriodic acid set free would react with sulphuric acid to give iodine and sulphur dioxide.

The Literature as a Guide in the Selection of Procedure

From what has been said above it appears that the selection of the best procedure may not be an easy matter even in the comparatively simple synthesis of a halide from an alcohol. It is true that many excellent procedures for individual cases are incorporated in laboratory manuals but, when there is a choice of procedures, or when there is no description of procedure, we must fall back on the method of trial and error. It would be wise, however, not to attempt such a method without consulting the literature dealing with similar problems. Supposing that the preparation of ethyl bromide from ethyl alcohol is a problem in procedure we would obtain, from the literature, the history given below.

Consulting the literature by following the directions given in Part IV, we find that, as early as 1827, a French chemist, Serullas (*Ann. Chim. Phys.* [2] **34**:99) had published an account of the preparation of ethyl bromide by treating ethyl alcohol with phosphorus and bromine. For some time this method was generally recommended for the preparation of ethyl bromide and many other alkyl halides and was illustrated in many textbooks—Roscoe and Schorlemmer, *A Treatise on Chemistry*

Preparation of Ethyl Bromide

(1901) III, 346: Kekule, *Lehrbuch der Organischen Chemie* (1861), 412: Erlenmeyer, *Lehrbuch der Organischen Chemie* (1883), 260. In 1857, de Vrij claimed (*J. Pharm.* [3], 31:169 and *Jahres-bericht*, 441) that a very good yield of ethyl bromide could be obtained by heating four parts of powdered potassium bromide with five parts of a mixture consisting of two parts sulphuric acid and one part 96 per cent alcohol (parts by weight is inferred from such descriptions). The mixture was heated until no more ethyl bromide came off, the vapors being condensed and led into a receiver. The receiver contained water. Ethyl bromide is insoluble in water: when no more drops of insoluble ethyl bromide descended through the water the distillation was at an end. It will be seen, on dividing the proportions given by de Vrij by the respective molecular weights, that the molecular proportion is .85 to 1 to 1. This corresponds only approximately to the equation $KBr + H_2SO_4 + C_2H_5OH \rightarrow KHSO_4 + C_2H_5Br + H_2O$. Apparently an excess of alcohol and sulphuric acid was considered desirable. Since this excess could not give rise to ethyl bromide it means that the yield would be something less than perfect. It is nevertheless a fact that, from the definition of yield, these proportions could give a 100 per cent yield (a so-called theoretical yield).

Yield

The yield is always expressed in terms of the percentage conversion of one particular substance into pure product and not in terms of all the reagents involved. The particular substance (unless otherwise noted) is taken to be that one of the starting materials which is most valuable of all the reagents employed, on account of the money or time or thought involved in its production. In the present case it is known that potassium bromide is far more expensive, mole for mole, than alcohol or sulphuric acid. Though an excess of alcohol and sulphuric acid was used it is obvious that we could get a 100 per cent yield if calculated from potassium bromide.

At the same time we may be fully confident that the excess of alcohol and sulphuric acid serves some purpose which is explained elsewhere in the literature.

Continuing the search, we find a recommendation to add $\frac{1}{4}$ to $\frac{1}{5}$ volume of water to the mixed materials (Dingler's *Polytech. Jour.* 229:284: Wagner's *Jahres-bericht* [1878], 538). This water is to be added carefully, with cooling, before distillation is begun (the cooling is understandable: the dilution of sulphuric acid generates much heat and is a slightly hazardous procedure). The volume of sulphuric acid, on the de Vrij proportions, would be about the same as that of the alcohol since it is over twice as dense (concentrated sulphuric acid of commerce is about 96 per cent sulphuric acid, density about 1.85: 96 per cent alcohol has a density about .80). The proportion of potassium bromide is also altered, a nearly theoretical yield of crude ethyl bromide being obtainable from 100 cc. of sulphuric acid, 100 cc. of alcohol, 50 cc. of water, and 90 g. of potassium bromide. Later experimenters slightly increase the proportion of water with no worse results. The addition of water apparently diminishes the amount of ether formed as a by-product in the reaction as well as the amount of free hydrobromic acid found in the distillate. The formation of ether (diethyl ether) in this reaction can easily be understood, the method generally used for the preparation of ether being to heat a mixture of ethyl alcohol and sulphuric acid to about 140°C. The temperature of the mixture given above, when distillation is proceeding at a fast rate, is about 120°C. and the amount of ether that is formed in this preparation is in consequence between 3 per cent and 5 per cent of the ethyl bromide. We can now see one good reason for using an excess of alcohol: we must allow for the unavoidable formation of ether. Also, since the reaction mixture is at 120°C., and alcohol boils at 80°C., and hydrobromic acid solutions boil at various temperatures up to 127°C. depending on the concentration of the acid, the detection of some ethyl alcohol and traces of hydrobromic acid in the distillate is not surprising. The use of an excess of sulphuric acid is still not explained. The suggestion for the use of water has been incorporated in most laboratory manuals, as, for example, Gattermann-Wieland, *Laboratory Methods* (1932), 81.

Weston (*J. Chem. Soc.* 107:1489) has described a procedure in which the formation of ether is almost entirely suppressed. The result depends on a slow distillation. The reaction mixture, as we are aware, is near 120°C. with a fast rate of distillation.

Two to three drops a second is a fast rate in the laboratory: the standard rate is about one drop a second.

The rate of distillation depends on the temperature of the mixture. Weston lowered the temperature of the mixture to the point where only a very slow distillation occurred and, by regulating the heating so that a very slow rate of distillation was maintained until the end of the reaction, he was able to avoid the formation of ether. This method does not seem to have been incorporated in laboratory manuals possibly because the time required would be longer than the usual laboratory period and because it is hardly worth while spending much time in the prevention of ether formation. The ether can be removed very simply from the crude ethyl bromide after the distillation (Friedlander, *Fortschrifte der*

Preparation of Ethyl Bromide

Teerfarb. 11:551. D. R. P. No. 52982). Weston obtained very high yields (90 per cent) of pure material and he did not use an excess of alcohol.

A slight excess of sulphuric acid was used: an explanation of the use of the excess sulphuric acid does not appear, however, and having collected all the needed information we are ready to decide on the amounts of the reagents. If we use the method involving a fast distillation, removing the ether from the crude ethyl bromide after distillation, the exercise can be completed in a few hours. Sodium bromide, in corresponding quantity, can be used in the place of potassium bromide. Calculate the amounts needed for a yield of 70 g. of ethyl bromide, the yield on the fast distillation method being about 85 per cent of pure ethyl bromide calculated from the potassium or sodium bromide.

Mixing

The materials can be mixed in the vessel used for the reaction. Sulphuric acid generates much heat on mixing with water: it is slightly hazardous to add water to sulphuric acid. For the control of the heat it would be best to mix slowly the alcohol and sulphuric acid, shaking and cooling under the tap. When cool the water can be added as ice. If ice is not available, add the water slowly, cooling well under the tap, and continue to shake, and cool well between the additions of portions of water. The powdered potassium or sodium bromide can then be added. The reaction, we know, does not start without heat.

When instructions call for the addition of one material to another it can be assumed that they are meant to be mixed uniformly by shaking or stirring: this is the operator's responsibility. It is sometimes dangerous to neglect this habit: reactions going forward with evolution of heat are sometimes so long delayed by the neglect of uniform mixing that enough of the material is introduced to generate an uncontrollable heat at some point. All reaction rates are increased at higher temperatures: the reaction may go forward with explosive violence (control of temperature, pp. **95, 119**).

We must arrange to heat the reaction mixture, to lead the volatile material through a condenser, and to collect the crude ethyl bromide (b.p. 39° C.). The distillation will be at an end when no more ethyl bromide distills over, that is, when no more insoluble drops of ethyl bromide will be seen descending through water in the receiver (descending because the density of ethyl bromide is about 1.4). If we were given no such convenient end point we would consider the use of a thermometer to tell the temperature of the vapors issuing from the reaction vessel, distilling until the rise in temperature of the thermometer gave some indication of the end point.

The Reaction Vessel

When distilling or boiling it is usual to employ a flask of a size about twice or three times the volume of the material as a safeguard against loss of liquid into the exit tube when sudden boiling occurs (distillation, pp. **14, 70, 72**).

The usual set of apparatus would give us a choice of (1) a long-necked round-bottomed flask, (2) a short-necked round-bottomed flask, (3) a conical (Erlenmeyer) flask, and (4) a distilling flask. With the first three flasks we would need a leading tube attached to the mouth of the flask and a condenser as illustrated in figure 6a (the bending of glass tubing is described on p. **60**).

The conical flask is usually reserved for operations in which it may be necessary to have access to some solid material inside the flask. It is structurally weaker than a round-bottomed flask: an evacuated conical flask, for instance, is liable to collapse under the pressure of the atmosphere. It is often used as a receptacle for materials being heated or boiled together (as in crystallizations) but hardly ever used for distillations. The round bottomed flask, preferred for the purpose because it allows more uniform heating of small quantities of material, is, as a rule, less liable to crack under heat stresses, and needs less adjustment in an assemblage of apparatus (distillation, pp. **14, 22, 24, 70, 107**). In comparing a long-necked flask and a short-necked flask there is this to be said: the long-necked flask allows more condensation of vapor in the neck in comparable circumstances. Some of the material that condenses in the neck flows back into the flask and this reflux condensation is an important factor in the method of separation of liquids by distillation. If, therefore, some separation of more volatile material from less volatile material is desired in the distillation from the flask we would prefer the long-necked flask (distillation, pp. **14, 22, 24, 70, 107**).

In the present instance (the flask contents at 120°C.) a long-necked flask will help in the retention of alcohol (b.p. 80°C.) and of hydrogen bromide (solutions of hydrogen bromide distill at various temperatures up to 127°C. depending on the concentration of the hydrogen bromide). The retention of hydrogen bromide should help to increase the yield.

If we use a distilling flask (the higher the side arm the more reflux condensation we achieve) there will be no need to fit it with a thermometer.

It is worth while mentioning that the short-necked flask has one feature which makes it very useful when firm connection is desirable: the sharp lip helps to prevent the slipping of rubber stoppers.

Preparation of Ethyl Bromide

Stoppers

Rubber stoppers are used when airtight connections are called for. They can be used as an alternative to cork stoppers in most operations in the laboratory, though cork stoppers are a little more resistant to the action of halogens and rubber stoppers are less liable to contaminate solutions of the alkalies. In ordinary laboratory practice a stopper is desired which will tolerate exposure to the vapors of boiling liquids for an hour or two without leakage or serious contamination of a distillate. For this purpose cork and rubber are both usable with most reagents (sulphuric acid, nitric acid, bromine are notable exceptions) but both lead to a slight contamination of many substances (many organic halogen compounds, high-boiling substances generally). Where the highest purity is wanted neither stopper should be used: they are both very complicated mixtures of organic and inorganic material some portion of which is likely to be soluble in any liquid we encounter. Cork stoppers are always rolled before use in order to soften them. When too dry, they can be softened by immersion in boiling water for a few minutes and subsequent rolling. If cork stoppers must be used in operations where air is to be excluded (as in some vacuum distillations), they are improved by first softening, then charring in a flame, and after placement, covering with collodion, or varnish, or rubber solutions. The slipping of rubber stoppers, mentioned before, is sometimes a serious matter. When exposed to the hot vapors of many organic materials (benzene, toluene, chloroform among common solvents) they are liable to swell after a time: unless firmly gripped in a sharp lip they are apt to slip out and, being swollen, it may be impossible to replace them in position. If a flask is very bluntly lipped, indeed, it is hardly safe to use rubber stoppers in it even for steam distillations or any distillations where slight strains and pressure tend to loosen the stopper.

Glass stoppers are used for the storage of organic compounds. They are sometimes liable to "freeze" in position under the corrosive action of some materials which attack the glass (alkalies are never stored in glass bottles for this reason: rubber stoppers are used). A tight glass stopper can usually be loosened by tapping with a piece of wood or metal. If not, then fixing the stopper in the jamb of door while the bottle is turned may lead to success. Certain devices are marketed for the purpose of loosening glass stoppers but the careful use of heat on the neck of the bottle outside the stopper is usually sufficient when the other methods fail. It is safer to use steam than a flame on the neck of the bottle: the repetition of the heating will often lead to success after the first attempt fails.

Bore stoppers from both ends to the middle. The borer should be kept wet and be turned without great pressure. It is difficult to bore a rubber stopper without diminishing the diameter of the hole at the centre of the stopper: use a little soap or alkali on a sharp cork borer and resist the temptation to hurry.

The Condenser

The leading tube (or side arm of the distilling flask) is attached to a condenser. The usual position of the condenser, arranged so that it is

Preparation of Ethyl Bromide

kept filled with running water is shown below. A few types of condenser are shown in figure 6 below.

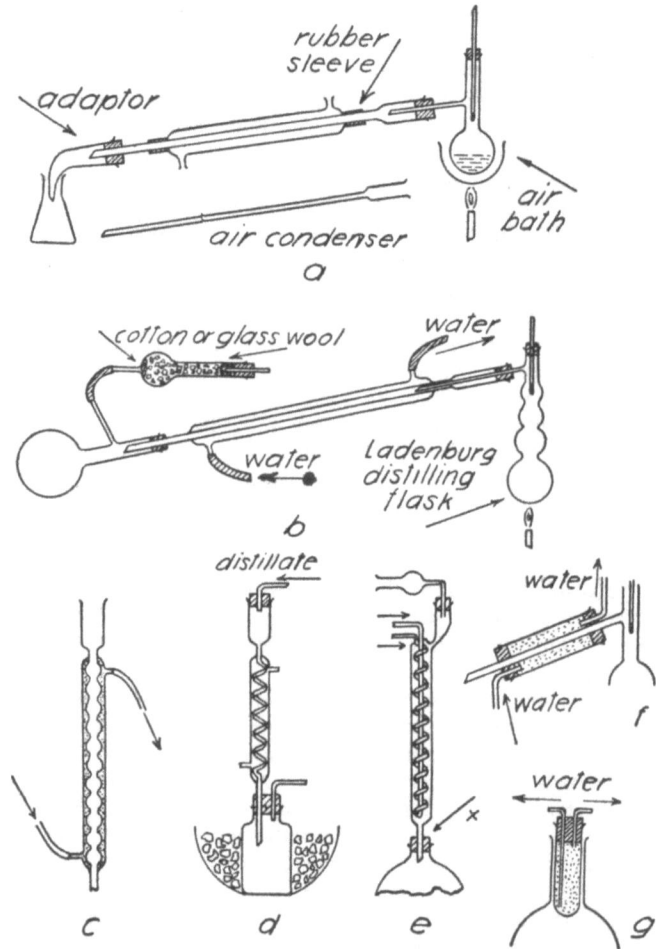

Fig. 6. (a) Liebig condenser and air condenser; (b) Liebig condenser made wholly of glass; (c) bulbed condenser for reflux condensation or downward distillation; (d) worm condenser for downward distillation only; (e) condenser with internal cooling; (f) small water condenser; (g) simple reflux arrangement for substances of high boiling point.

6a is the ordinary Liebig condenser (it is made in various lengths) with rubber attachment of inner tube to outer jacket. When good cooling is re-

Preparation of Ethyl Bromide

quired, as in steam distillations at a fast rate or with very volatile material such as ether and ethyl bromide, the condenser should be about 24" in length (or two small condensers should be attached to each other). Liebig condensers are available in one piece (figure 6b) but they are liable to crack at the glass joint unless the leading tube is extended, as shown, past the shoulder of the condenser. This limits the size of leading tube that can be used with them; even in the best position of the leading tube they are liable to crack at one or other of the glass joints if distillation is rapid (particularly in steam distillations where the high specific heat of the steam is a factor).

Figure 6c shows a bulbed condenser which gives a greater condensing surface than the Liebig condenser for equal length of tubing, and is best used in the upright position shown. In the inclined position, material collects in the tubes. Such material can be run out of the tubes at the end of distillation, of course, but in most distillations certain portions of distillate should not be mixed with later portions: the prevention of such mixing in the bulb could only be accomplished by detaching the condenser and draining it whenever a fraction of distillate is cut off.

A worm condenser (figure 6d) is used only in the upright position. This condenser, again, gives great cooling surface per unit length. It has the drawback of being partly filled with condensate when distillation is rapid, causing mixture of portions of distillate which, perhaps, should not be mixed.

Figure 6e shows a very efficient condenser which can be used in the inclined or upright position and which has the merit of internal cooling. When the cooling is external the cold outer surface of glass is liable to cause deposition of moisture on humid days. This moisture may collect into drops and run through a leaking cork into a reaction mixture or into a distillate. The internal cooling eliminates the trouble of disposal of the deposited moisture.

A simple small condenser, such as can be made in a few minutes, is shown in figure 6f.

All the condensers mentioned above can be used as reflux condensers in the upright position shown in figure 6c, condensed liquid flowing back into the flask. Figure 6g shows a simple arrangement for refluxing liquids of fairly high boiling points—100°C. or over (condensation, pp. **28, 64, 74**).

The Receiver

The material to be collected boils at 39°C. The exposure of such volatile material to the atmosphere for an hour or two in the ordinary bottle or flask would result in the loss of a few grammes of yield through evaporation. The usual receiver is a conical flask, an adaptor being placed at the end of the condenser (figure 6a): it would be well to arrange to keep the receiver cool in this exercise. Since we are to employ water to mark the end of the distillation, we can use ice in the water to aid in cooling: the cooling will be more effective if the distillate enters directly into the water rather than into the atmosphere above the water (the heat conductivity of water is much greater than that of air). When the adaptor lip is below the surface of the water, however, there is some danger, at some time or another, of a backflow of water into the condenser and flask. Heat fluctuations, or solubility of some of the vapors in water, can cause such a backflow. To safeguard against this, either the adaptor must be arranged so that it can easily be tilted above the surface of the water or the connection between condenser and adaptor should be easily broken (a loose-fitting cork stopper would be suitable).

The amount of water needed in the receiver is the next matter for attention. In the distillation of the ethyl bromide, the excess ether and alcohol come over with the ethyl bromide in the distillate. For the receiver, then, we need enough water, only, to make certain that the end point will not be obscured, or any material lost, by solubility in the water or in the mixture of water and crude distillate. The distillate will contain alcohol (soluble in water), ether (very slightly soluble in water), and ethyl bromide (insoluble in water). Ethyl bromide is soluble, however, in alcohol: it will dissolve to some extent, therefore, in mixtures of alcohol and water. If we assume the worst, that all excess alcohol will be found in the distillate, we must use enough water to be certain that the mixture of alcohol and water will be so dilute in alcohol (say not more than 10 per cent) that the ethyl bromide will be practically insoluble therein (extraction, pp. 12, 13, 30, 40, 72). If the water is too sparingly used it may result in a large loss of ethyl bromide by solubility in the alcohol-water mixture. The receiver suggested here has the disadvantage of requiring attention: it may require attention at some moment when attention is urgently needed elsewhere. Such apparatus, which may need attention at two points simultaneously, is not to be generally recommended. A bottle surrounded by ice and water would need no attention even though the level of the distillate increases (figure 6d). Using a deep bottle the distillate could drop into a layer of water for marking the end point. Since, however, the cooling effect of air is not so great as that of water (by reason of its smaller heat conductivity), to aid the cooling it would be well in this case to use a long condenser or two small condensers and also to cover the bottle loosely with cotton wool.

Heating

Figure 6a shows an air bath in position for heating. The air bath is merely a bowl of metal surrounding the lower half of the flask and separated from it by an air space of about half an inch. It gives more uniform heating than can be achieved with a direct steady flame but is hardly better than a moving flame. A sand bath (a layer of sand in a bowl of metal) would give more uniform heating over the surface of contact and could be used as an alterna-

Preparation of Ethyl Bromide

tive: in this case the flask may rest on the sand and may be clamped very loosely or not clamped at all. One of the foregoing baths is suggested in this exercise simply because the mixture dealt with is rather viscous and is more liable to severe bumping by superheating than less viscous material would usually be. An oil bath is sometimes useful, the temperature of the bath being observed with a thermometer. Paraffin wax, rape seed oil, or "arochlor" can be used in such baths: they are usually unpleasant at temperatures above 220°C. A metal bath (of low melting alloys such as Wood's metal) is useful for higher temperatures.

In this exercise the bath is heated with a fairly strong blue flame until distillation starts and, for the sake of speed, the distillation is kept up at a fairly fast rate (2 to 3 drops a second) until no more ethyl bromide comes over. The flame is adjusted to keep up the fast rate of distillation: do not heat so strongly as to cause froth to run into the condenser.

Towards the end of a distillation it is usual to raise the temperature of the heat source a little in order to make certain that the end point is not a fictitious one: an unnoticed draught of air or a lowering of gas pressure could lead to a cessation of distillation entirely and thus deceive the operator—whose attention was concentrated on the distillate—into believing that an end point has been reached (distillation, pp. **14, 24, 70, 107**).

Purification

The method recommended for removing the ether and alcohol in the ethyl bromide is that of shaking with cold concentrated sulphuric acid (D. R. P. No. 52982). The lower layer of ethyl bromide is run off from the distillate by the use of a separatory funnel. If a large excess of alcohol was used in the original mixture it would be well to wash with water once so as to remove some of it. If any suspicion exists that the upper layer may contain an appreciable quantity of ethyl bromide the upper layer should be diluted with ice and water: any bromide settling out should be mixed with the main portion. The instructions for removing the ether and alcohol simply advise mixing cold concentrated sulphuric acid into the crude bromide (which is kept cold in a freezing mixture) and using enough sulphuric acid so that, even when diluted with ether and alcohol, it is still sufficiently concentrated to be denser than ethyl bromide: that is, about 70 per cent sulphuric acid even when diluted. It should be kept quite cold so as to avoid reaction between the bromide and the acid. The sulphuric acid should be kept concentrated because ether and alcohol are not soluble in dilute sulphuric acid. It appears that one treatment is enough, adding sulphuric acid drop by drop till a layer of sulphuric acid appears below the ethyl bromide.

From our knowledge of the partition of material between two solvents, we would not expect the alcohol and ether to be removed completely in one treatment from the ethyl bromide in which they are infinitely soluble to the sulphuric acid (even if they are infinitely soluble in sulphuric acid). It must be a matter not involving mere solubility: there is, as a matter of fact, compound formation between the ether (or alcohol) and sulphuric acid. The compound will be referred to later: here we only note that it must be a compound that is insoluble in ethyl bromide and quite soluble in concentrated sulphuric acid (extraction, pp. **12, 28, 40, 72**).

With these instructions we should be apparently correct in placing the crude ethyl bromide in a conical flask (a beaker is more liable to lead to loss of bromide by evaporation) and surrounding the flask with a freezing mixture. The sulphuric acid could then be added slowly from a dropping funnel and the mixture swirled or stirred with a rod (swirling would give better mixing of the liquids). We could add sulphuric acid until a lower layer of sulphuric acid became evident, bearing in mind that the first portions of acid, greatly diluted by alcohol and ether, may form a layer above, and not below, the ethyl bromide.

Cooling

The temperature of an ice-water mixture can be lowered by the addition of common salt: the rate of cooling is thereby increased. The freezing mixture of ice and salt is most effective when the ice is finely ground, the salt thoroughly mixed in, and the mixture kept stirred occasionally so as to prevent caking. With salt to the extent of about one-third of the weight of ice, a temperature of $-21°C.$ may be reached. Hydrated calcium chloride and crushed ice (or snow) in the ratio of 10 parts to 7 parts by weight give a temperature of $-54°C.$ Solid carbon dioxide in ether, chloroform, acetone, kerosene, or other low-freezing liquids will give even lower temperatures (about $-110°C.$ with ether), but such mixtures have a somewhat low heat conductivity which impairs their usefulness as a freezing mixture in the ordinary work of the organic laboratory where, as a rule, much heat is to be conducted away. The heat conductivities of water, ether (and most organic liquids), and air are worth comparison: the ratio is about 100 to 25 to 4. This last figure accounts for the advice already given to cool ethyl bromide in ice water: it also accounts for the practice of retaining a little water or salt solution in the usual ice-salt freezing mixture. Dry freezing mixtures have numerous air spaces in contact with the surface to be cooled. In practice, then, an ice-salt freezing mixture is made of chopped or crushed ice, well mixed with salt. A little salt solution is retained so as to cover the bottom of the receptacle being cooled: the excess solution is siphoned away. In order to provide room for the work of mixing and adding fresh ice as desired, the mixture is held in a dish or a very wide beaker. Reserves of ice are kept handy and when a temperature near $0°C.$ must be maintained for some hours, or when a strongly exothermic reaction is to be controlled, the weight of the freezing mixture should be at least four times the weight of the material to be cooled. The rate of cooling will depend, also, on the

Preparation of Ethyl Bromide

amount of surface exposed to the freezing mixture so that, for careful control, the material is best kept in a receptacle providing the greatest amount of cooling surface for unit volume of material (small depth or small cross section of liquid). In the present case we need only a mediocre freezing mixture (cooling, pp. **30, 95**).

When the lower layer of concentrated sulphuric acid has become evident it should be carefully run off in the separating funnel and the ethyl bromide transferred to a distilling flask ready for distillation. In order to prevent the bromide mixing with traces of sulphuric acid left in the tube of the separating funnel below the stopcock, it should be poured from the top of the funnel into the distilling flask. This bromide, already dried by the sulphuric acid, can be distilled without further treatment: with most materials, of higher boiling point, it would be necessary to wash with sodium carbonate solution at this point to remove traces of sulphuric acid (washing, pp. **10, 36, 59, 63**). Choose the type of distilling flask which gives a maximum of refluxing surface (high sidearm or the Ladenburg flask illustrated in figure 6b) and collect the material in a dry glass-stoppered bottle held in ice water.

As the temperature of the material in the bottle rises to that of the room, an expansive pressure is exerted by the vapor of the bromide: release this pressure occasionally until room temperature is reached by removing the stopper and then stopper firmly for storage: the yield should be noted (stoppers, p. **25**).

Distillation

The normal rate of distillation in the laboratory is about one drop a second. This is the rate used when finding the boiling point of a liquid. The thermometer is kept in the position shown in figure 6a, and not in the liquid itself. We observe in this way the condensing point of the vapor which is the same as the boiling point of the liquid. When a pure liquid is being distilled it will be noticed that the temperature on the thermometer rises very rapidly to a point a few degrees below the boiling point and that, after a few drops have passed over, the thermometer remains steady at the boiling point. With a steady rate of distillation the recorded temperature remains constant: fluctuations in the rate of distillation may cause a fluctuation of a degree or so in the thermometer. Towards the end of the distillation, with only a few cc. left in the flask, a rise of 2°C. to 3°C. may be caused by superheating which is difficult to avoid at this stage. In the purification of liquids by distillation, the end is achieved when the behavior is that described above and the behavior must be that of the whole of the material undergoing distillation. It is not sufficient to take a certain constant boiling fraction of some impure material: such fraction must be redistilled, and, if necessary, redistilled again, until we have a portion which behaves as a pure liquid when totally distilled. Then, and only then, the product can be labeled b.p. 38°C.-40°C., or whatever the observed range is, most of the so-called "pure" organic compounds having a boiling range of ½°C. to 2°C. Since the boiling point is the point at which the vapor pressure of the liquid is equal to that of its atmosphere, we may observe different boiling points with the same liquid at different times because of variation in the atmospheric pressure. These barometric variations may change the boiling point two or three degrees (distillation, pp. 14, 24, 70, 72, 84).

Constant Boiling Mixtures

Certain mixtures show the behavior described above for pure liquids, that is, they distill totally at a constant temperature. Such mixtures are not very common. Constant boiling mixtures of inorganic compounds have probably already been encountered. (A solution of hydrobromic acid in water (47 per cent) or hydrochloric acid in water (20 per cent) distills completely at a constant temperature). Some classes of organic compounds, however, are notorious for the prevalence of constant boiling mixtures (hydrocarbons with ether and alcohols with water, for example). When such a case is suspected it can usually be confirmed by a distillation at a different pressure. At such new pressure, a pure liquid would again distill at a constant temperature whereas the mixture which distilled at a constant temperature at the original pressure would now distill over an extended range of temperature and so demonstrate its true nature; this because the proportion of constituents which makes up a constant boiling mixture at one pressure is not necessarily, nor even likely to be, the same proportion as that which makes a constant boiling mixture at a new pressure. The proportions of constituents, furthermore, would be hardly likely to coincide with an exact stoichiometric proportion. Hydrochloric acid (20 per cent solution) was long considered a compound because the proportion of acid to water was almost exactly 1 mole to 8 moles but it was later observed that the proportion varied with the pressure under which the distillation was performed. These constant boiling mixtures have either a lower vapor pressure around the boiling temperature than either constituent alone (as with 20 per cent hydrochloric acid) and so show a boiling point above either constituent, or they have a higher vapor pressure than either constituent alone (96 per cent alcohol boils at a lower temperature than alcohol or water or any other mixture of water and alcohol). In attempting purification of the wanted material by distillation, therefore, repeated distillation with a view to isolation of constant boiling material (see fractional distillation) merely leads to the isolation of the constant boiling mixture.

THE LABORATORY NOTE BOOK

The instructions in this experiment have been deliberately made obscure. Set up in your notebook the equations, amounts of material, detail of conduct of the experiment, times, and precautions. State what apparatus you will use. Put aside this laboratory manual and get the approval of the instructor for your abstract of the procedure. Enter all relevant matter and compare your yield with neighboring workers. Be prepared for questions and to question.

Exercise 3

Purification of a Known Substance by Crystallization
(17, 22, 36, 56, 80)

THIS EXERCISE illustrates the usual procedure for purification, by crystallization, of a solid of known properties. The impure substance, in this case acetanilide, anisic acid, benzoic acid or salicylic acid is contaminated with impurities of unknown nature.

A solvent is sought which is the nearest to the ideal. An ideal solvent for the purpose would be one which dissolves a large percentage of the given material when hot, a negligible percentage when cold. It would either not dissolve the accompanying impurities, in which case the impurities could be filtered or otherwise removed from the hot solution, or it would retain the accompanying purities in solution when cooled. It must not react with the material and it should be easily removed from the material after crystallization (crystallization, pp. **37, 162**).

Given such a solvent the process of crystallization would be merely that of solution in the hot solvent, usually at the boiling point, filtering any undissolved impurities, and allowing to cool. When the crystals are deposited they are filtered and washed with a little of the same solvent—to remove the adhering original solution containing impurities—and then dried.

In the present case, an examination of solubility tables in handbooks, in Seidell, *Solubility of Organic Compounds*, or in the original literature as shown in Part IV, shows that the solubilities of these substances are as given below in three common solvents. The solubility is expressed as weight of substance in 100 g. of solvent.

	ETHYL ALCOHOL		WATER		BENZENE	
	Boiling	*Cold*	*Boiling*	*Cold*	*Boiling*	*Cold*
Acetanilide....	> 150 g.	20 g.	6 g.	0.5 g.	> 10 g.	1.0 g.
Anisic acid....	> 150 g.	53 g.	> 5 g.	0.1 g.	....	....
Benzoic acid...	> 150 g.	50 g.	6 g.	0.2 g.	> 50 g.	7.0 g.
Salicylic acid ..	> 150 g.	50 g.	7 g.	0.2 g.	> 10 g.	1.0 g.

We do not know the nature of the accompanying impurities so that we are unable to decide absolutely which solvent would be preferable on the grounds of solvent effect on impurities. It is known, however, that water does not dissolve most organic compounds and, on the assumption that the impurities are mostly organic, water would be preferred in the hope that the accompanying impurities would not dissolve in it to any appreciable extent. The solubilities in water are also very convenient for working with the usual amount in laboratory manipulation (10 g.- 100 g.).

Crystallizations are most conveniently done from boiling saturated solutions. The solvent chosen should therefore be preferably one whose boiling point is below the melting point of the substance to be crystallized so that the material coming out of the hot saturated solution on slight cooling will be crystalline and not molten drops of the material. If the best solvent on other grounds is, however, such a solvent whose boiling point is above the melting point of the material, it is used with a slight modification of the usual technique (solvents, pp. **40, 101, 155, 162**).

It will be seen that the melting points of the substances given above are all above the boiling point of water so that water should be satisfactory for the ordinary operation outlined below.

Weigh the crude material. Assuming it be almost pure material, introduce it into a beaker containing a 5 to 10 per cent excess of water for complete solubility when boiling. Bring to the boil and boil for a few minutes with stirring.

If the solvent chosen were any other than water the beaker would not be used—a conical flask under a reflux condenser would be the standard equipment because organic solvents are either too expensive, or too inflammable, or too poisonous, to be heated in a beaker in this way (crystallizations, pp. **101, 155, 162**).

During the boiling, some water will be lost by evaporation, but little or none of the material.

As a general rule we can assume that substances which are solid at room temperatures and which are totally dissolved in a solvent boiling below or near 100°C. will not be removed to any appreciable degree by the evaporation of a small fraction of the solvent. This would not be a safe assumption if the solid were insoluble in the solvent, particularly if the solvent were water (for reasons see pp. **44, 45, 84, 107**).

The wanted material being now in solution, any other material not in solution must be removed. If a solid material remains undissolved it can be removed by filtering.

For the filtration of the hot solution the jacketed funnel shown below could be used. This funnel, which is kept hot with a flame, is unsuitable for any solvents but water, however, and for the hot filtration of organic solvents a steam-line jacket of copper tube wound in conical form (C) could be used as an alternative.

It is nearly always the case that the use of such funnels can be avoided if

Purification by Crystallization

the ordinary Buchner funnel and the thick filter flask are first warmed (with steam or hot water or in an oven) and then, with the aspirator pump running at full speed, the hot solution is brought on to the warm Buchner funnel as quickly as possible. With a slight excess of solvent the solution is usually filtered rapidly enough to prevent the lowering of the temperature of the solution to the point where crystals of the material are deposited above the filter or in the pores of the filter paper. If the filtration is not

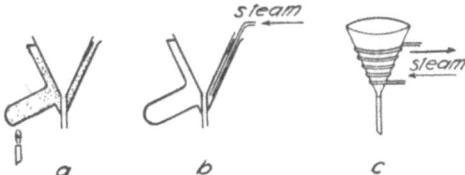

Fig. 7. (a) Hot water jacket; (b) hot water jacket used as a steam jacket; (c) steam coil jacket.

completed within 30 to 60 seconds it is usually hopeless to expect success by this method. If the hot jackets are not available and the solution of material cannot be filtered by the Buchner when saturated, a more dilute solution must be made and the excess solvent evaporated after the filtration (filtration, pp. 10, 36, 105).

Though the substances listed above are colorless, it may happen that, on bringing the solution to the boil, the solution appears colored or tarry with impurities. If such is the case the solution is treated with a little charcoal and boiled for some time in the hope that the colored material will be adsorbed on the charcoal and so removed when the charcoal is filtered off.

Animal charcoal, or some prepared charcoal, is generally used for the removal of colored and tarry impurities from solutions. It is used sparingly (usually in the first trial in no greater amount than 5 per cent of the weight of the material sought) and is kept in contact with the solution, usually at the boiling point, for some time. Often enough the decolorization is effective in a few minutes but it sometimes requires hours. As charcoal is capable of adsorption of most organic material, it is used sparingly until it is established by trial that larger quantities are necessary and not detrimental to the yield.

The effectiveness of charcoal depends on its property of preferential adsorption of substances of high molecular weight. Usually the tarry or colored impurities in solutions of organic substances are of high molecular weight. When this is not the case, the removal of the color may be impossible by this method without too large loss of material.

On bringing the solution to the boil it may happen that liquid drops of some impurity can be seen. If they are denser than the solution they

Purification by Crystallization

can perhaps be removed by decantation. If they are less dense than the solution they may be removable in great part by the careful use of filter paper passing over the surface and finally by filtering the hot solution through an earth (Kieselguhr, Fuller's earth, etc.).

The bed of earth is prepared by mixing with water or the solvent being used and filtering on a Buchner funnel. The bed is washed well with the hot solvent (filtration, pp. **10, 34, 105**).

The hot filtered solution, which has been treated with charcoal if colored or tarry, is quickly poured into a conical flask or beaker and allowed to cool. While a beaker is permissible in this exercise it is not generally advised.

A conical flask is used for most crystallizations and is usually stoppered to prevent loss of solvent (crystallization, pp. **101, 155, 162**).

When the crystals appear and the solution is cold, the specimen is filtered, washed with some of the same solvent (in this case water), and pressed with a glass stopper. The amount of solvent to be used in washing will depend on the solubility of the material in the solvent. The pump is disconnected after pressing and the material just covered with a little of the solvent and stirred with a glass rod so that it is distributed in the solvent. The pump is again connected and the mass pressed after the solvent has been drawn off. Usually this process is again repeated twice.

On cooling the solution it may also happen that the substance first comes out as oily drops though the temperature is below the melting point of the substance and these oily drops later solidify. If such a thing occurs it is good practice to stir very vigorously until crystals appear. This first appearance in the form of oily drops could be occasioned by the presence of an impurity which comes out with the material (such impurity causing a lowering of the melting point as discussed below) but it could occur with pure material (acetanilide for example) for a reason discussed later (see p. **162**). It is good practice, in any case, to stir vigorously until crystals appear.

The crystals so obtained could be tested for purity on a corrected thermometer by comparing their melting point, after drying, with the melting point given for that substance in the literature (acetanilide 114°C., anisic acid 184°C., benzoic acid 122°C., salicylic acid 159°C.). A more informative practice, however, is to crystallize the material repeatedly until the melting point shows no further change upward: i.e., until two specimens from successive crystallizations show the same highest melting point when observed simultaneously in the melting point apparatus. This practice, common in the examination of new compounds, allows us to defer the matter of correction of a thermometer till later. In

Purification by Crystallization

such practice a small quantity (about ½ g.) of the crude material is set aside as a specimen and the same is done with each crop of crystals. The first crop of crystals, diminished by the removal of a specimen, is recrystallized using the proper proportion of water—this is less than in the first crystallization (calculate). If charcoal or earth is again found necessary it should be employed. A specimen is again removed (½ g.) and crystallization repeated until crops of pure white crystals have been obtained twice in succession. The melting points of the specimens are observed as shown below. If the melting points of the last two specimens are the highest of all and are exactly the same when observed simultaneously the crystallization is at an end.

Pure substances which are stable in the temperature region of the melting point have a sharp melting point. In the apparatus shown below, which is more convenient than accurate for the estimation of melting point, this appears as a complete melt (first softening to clear melt) in ½° C. to 1° C. when the heating is conducted in the ordinary way. Impurities lower the melting point (pp. 36, 105, 151) and also extend the range of melting so that purification is accompanied by a raising and a shortening of the melting range which is always described as covering the temperature from first softening to complete clear melt.

If the specimens show such a transition to higher and sharper melting points until two crops are identical, we are able to state that the solvent purifies the material to a certain degree (which is perhaps pure, and which is likely to be pure if the final melting point is sharp) and that further use of the same solvents, even if the crystals are not pure, is not warranted (crystallization, pp. 105, 151, 162).

The melting point of the final specimens in this exercise, which we can assume to be pure, will be found to lie near the melting point quoted from the literature. Deviations from the quoted melting point will not be exactly the same with all thermometers, however, since the thermometers are not standardized. Even with a standardized thermometer the final melting point observed with this apparatus is likely to be a degree or two low because the quoted temperatures are "corrected temperatures": that is, those obtained when a standardized thermometer is used correctly with the column of mercury totally below the surface of the bath liquid or when a standardized thermometer has been used in some such apparatus as the present one and the melting point corrected, by a formula, for the portion of mercury column not immersed (melting point, pp. 36, 98, 106).

First dry the specimens. They can be dried by placing in an air oven, spread out on hard paper, tile, or glass, and protected from the metal

38 Purification by Crystallization

plates of the oven with a thick asbestos pad, for an hour or two at 80°C.-90°C.

In general, the temperature of the oven should be near the boiling point of the solvent from which the material to be dried has last been crystallized, and it is not always true that a few hours will suffice. This temperature should be at least 20°C. below the melting point of the pure material and, when there is any fear of melting, the specimen should be spread on a glass container such as a watch glass, in order to prevent loss. Occasional pressing with a spatula often expedites the drying. When the melting point of the substance is in the region of temperature desired for drying, then a vacuum desiccator or an ordinary desiccator can be used, but in this case the time given must be much longer. When time presses, the material can sometimes be treated in the following way: it is washed on the filter paper by a volatile solvent which will dissolve away the original solvent without too great loss of the material (sometimes the volatile solvent is again washed away with a more volatile solvent) and is then dried merely by exposure to air with occasional pressing. Thus high-boiling hydrocarbons are often washed away with benzene, or acetic acid with water and the water with alcohol and the alcohol, if necessary, with ether (drying, pp. **38, 50, 72, 82**).

There will be no difficulty with these materials if dried at 80°C.-90°C. in the oven, or if they are washed (very carefully) with a very small quantity of alcohol and dried by exposure to the air or by pressing repeatedly on porous tile.

Using a spatula, powder a small quantity of each dry specimen on separate pieces of porous tile (3 to 4 sq. in. in each piece) or hard paper. If the crystals are too hard to be powdered in this way, then a mortar and pestle should be used. Introduce a little of the specimen into the capillary tube, scratching the capillary tube gently with a file so that particles adhering to the side fall to the bottom. Continue until the depth of material in the capillary tube is about $\frac{1}{4}''$ after packing the mass together by tapping the tube a few times on the bench. The capillary tube is made from glass tubing ($\frac{1}{4}''$ or more in width) or from small test tubes.

These are quickly heated in a hot blue flame of a wing top-burner or blast lamp with continuous rotation until partly collapsed and quite soft. The glass tubing or test tube should then be drawn out with a pull of steadily increasing power until the extended portion has an internal diameter of about one millimeter. Suitable pieces about 4″ in length are then broken off. The pieces chosen for capillary tubes should be of nearly the same internal diameter and thickness and the user should work always with capillary tubes of this chosen dimension. One end of each capillary tube is fused to a small clear blob of glass by quick rotation in the edge of a hot blue flame; it is more important that a clear fused spherical blob be made than that the fusing should look neat, since there is danger of a minute hole being left

Purification by Crystallization

leading to an incorrect melting point by entrance of bath liquid into the capillary tube (glassworking, pp. **60, 61**).

Wet the bulb of the thermometer in the sulphuric acid of the melting-point apparatus and place the capillary tube or tubes in the positions

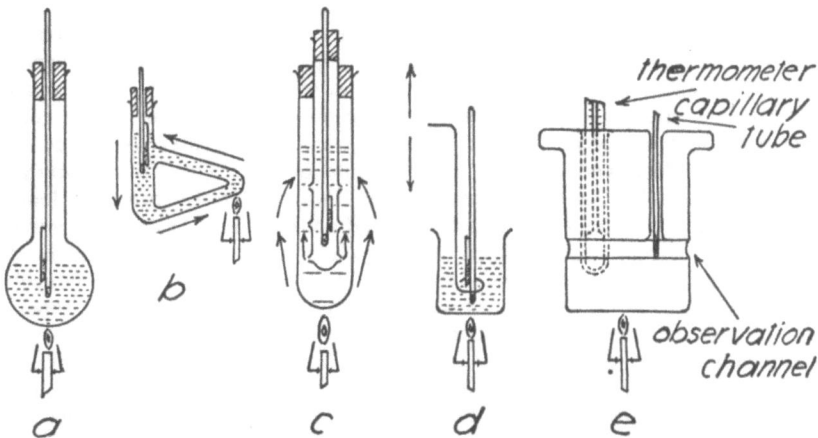

Fig. 8. (a) Simple melting point apparatus (in this apparatus and in (b) and (c) the corks are grooved so as to enable the thermometer to be seen); (b) Thiele melting point tube; (c) Fisher melting point apparatus with capillary tube too short; (d) melting point apparatus with stirrer; (e) copper melting point block.

shown: they will adhere to the thermometer on wetting with the acid. Carefully replace the thermometer in position and employ capillary tubes of such length, and sulphuric acid in such amount, that about an inch of capillary tube is left above the surface of the sulphuric acid. If such a length is not left for capillary attraction of the tube to the thermometer the capillary tube will fall away from the thermometer. If the bath liquid is such that the capillary attraction is small, it may be necessary to attach the tube to the thermometer with a rubber band.

Heat with a small flame protected by a chimney, the chimney top being within ½ inch of the outer tube of the apparatus for best protection against air draughts. Bring the temperature to within twenty degrees of the melting point and then carefully adjust the flame so that the temperature rises at a steady rate of 2°C., 3°C. or 4°C. a minute; let the chosen rate be the standard for your future determinations, as also the height you have chosen for the bath liquid, the position of the thermometer, etc., etc. Observe the melting point, recording it as the interval between the temperature of distinct softening and the temperature of clear complete melt.

Adverting once more to the matter of solvents, it will be recalled that the one chosen here for crystallization was used in the hope of leaving impurities undissolved. This total process might be called an extraction of material away from impurities and subsequent crystallization. In dealing, as is common, with impurities of unknown nature and material of unknown solubilities, it is often necessary to make tests of comparative efficiency of solvents. A small quantity of the material is placed in a test tube or, better, in a small conical flask under a reflux condenser which may be an ordinary water condenser or, if the solvents boil above 150°C., an air condenser. A few cc. of solvent are added and refluxed for a minute or two. The flame is extinguished and the hot solution either decanted away or filtered through a loose plug of cotton wool in an ordinary filter and collected in a beaker. By rough comparison of residues it can be decided which solvent has taken up most material. By rough comparison of the amount of crystals coming out in each case a decision can be made as to which solvent gives up the greater proportion of dissolved material on cooling. If the melting points are now observed a decision can be made, from the range and height of the melting point, on the question of the comparative efficacy of the solvents in removing impurities. With all results in mind the best solvent is chosen (solvents, pp. **33, 40, 72, 102, 155**).

Exercise 4

Preparation of Chloroform

(23, 27, 35, 79)

THE PRESENT chapter is devoted to the discussion of a procedure for the preparation of chloroform on a small scale in the laboratory. Text books on organic chemistry state that chloroform can be obtained from various source materials and in various ways. It is evident, however, that only the "haloform" reaction has been much used in preparative work. In this reaction a compound containing the grouping $CH_3CH(OH)C-$ or CH_3COC- or simply $CH_3CH(OH)H$ (ethyl alcohol) or $CH_3CO(H)$ acetaldehyde, is subjected to the action of a halogen in the presence of an alkali. The hydroxyl group, if present, is oxidized to the carbonyl group and substitution of the halogen occurs on the methyl group next to the carbonyl group. This substitution of the halogen is followed by scission under the influence of the alkali into chloroform (bromoform, or iodoform) and an acid. Thus $RCOCH_3 + 3NaOCl \rightarrow RCOONa + CHCl_3 + 2NaOH$. The reaction, obviously, can be used for the preparation of an acid having one carbon less than the original methyl ketone. It is often so used. In the preparation of chloroform itself it is natural that attention should be directed to the most readily accessible materials as reagents in the preparation. The two materials commonly employed are ethyl alcohol and acetone, both cheap, readily accessible, and stable. Acetaldehyde is neither so cheap nor so stable as the other two source materials.

The Literature

Chloroform was first described by Soubeiran (*Ann. Chim. Phys.* [2], **48**:131). It was prepared from alcohol by the action of bleaching powder. Liebig also described its preparation by the same method (*Ann.* 1:199) and mentioned the fact that it could be obtained in better yield by the use of acetone (essiggeist). Nevertheless most laboratory manuals up to 1890 recommended alcohol as a starting material though by that time many manufacturers had come to prefer the more expensive acetone even for large scale work. This preference for alcohol was possibly an outcome of the report of Siemerling (*Arch. for Pharm.* [2], **54**:32), that acetone, in addition to being more expensive, gave a less satisfactory yield than did ethyl alcohol. Orndorff and Jessel (*Am. Chem. J.* 10:365) compared carefully acetone and ethyl alcohol as starting materials and came to the conclusion that acetone was far more satisfactory for laboratory work. The yields were higher and the operation was less troublesome. This is the method that will be discussed in detail below. Since chloroform is a very useful material in medicine and chemistry it is not surprising to find that much attention has been paid to the details of manufacture. It is apparently possible to get, by the alcohol method, yields of chloroform approximately equal to the original weight of the alcohol. The acetone method presumably gives yields approximating 160 per cent of the original weight of the acetone with the same grade of bleaching powder (Orndorff, *loc cit.*).

Procedure

The method of preparation from acetone and bleaching powder involves, like the original method, much frothing. The reaction is exothermic and proceeds rapidly even when the acetone is greatly diluted or the bleaching powder suspended in a large volume of water. The physical properties of the reagents and products are the following. Acetone is a liquid boiling at 56°C., density 0.8, miscible with organic solvents and with water. Chloroform is a liquid, density 1.5, boiling at 61°C., miscible with organic solvents and nearly immiscible with water (solubility in water 0.8% at 20°C.). Bleaching powder approximates in composition $CaOCl_2$, is only slightly soluble in water, and one of the major reaction products, calcium hydroxide, is also only slightly soluble in water. The reaction mixture, therefore, will not be homogeneous at any stage.

According to Orndorff, the reaction corresponds to the equation $2CH_3\text{-}COCH_3 + 6CaOCl_2 \rightarrow 2CHCl_3 + 3CaCl_2 + 2Ca(OH)_2 + (CH_3COO)_2Ca$. Bleaching powder, however, is a trade product which varies in composition and is unstable to atmospheric moisture and carbon dioxide. It is specified in terms of its "available chlorine" and we should use about twelve parts of a good grade of bleaching powder (available chlorine from 30 per cent to 35 per cent) to one part of acetone if we follow the procedure of Orndorff. If bleaching powder were actually represented as in the given equation we would have needed only about six parts. Fresh commercial bleaching powder can be considered to have about 33 per cent available chlorine. Setting before ourselves the problem of converting in the most economical way some 15 g. of acetone (or 19 cc.) into chloroform, we must take account of the physical

Preparation of Chloroform

properties of the materials, the frothing, the heat of the reaction and so on.

Frothing

The simplest way to deal with frothing is to allow sufficient space in the reaction vessel for the foam to break before it reaches any point leading out of the vessel. This, of course, will be a matter of trial. If the frothing occurs, as here, during a reaction, it may be controlled to some extent by allowing the reaction to proceed slowly either by governing the temperature or by slow mixing of the reagents. It will be found that the frothing can be checked in the present instance by the use of a larger flask than usual (four or five times the volume of the reaction mixture) and by slow mixing of the reagents. More difficult cases often occur: distillations and evaporations are frequently complicated by persistent frothing which tends to contaminate the distillate or lead to loss of material. In the organic laboratory the addition of a little ether or amyl or heptyl alcohol is often recommended for the purpose of breaking a foam but no general argument can be given in the matter (see text books on colloid chemistry and engineering chemistry for detailed discussions). In all cases we would try to avoid the use of added materials since these bring their own problems of separation in their train. The use of an obstructive baffle, glass beads, or a current of cold air is sometimes sufficient to break a froth.

For maximum efficiency in the conversion of acetone to chloroform, it will be necessary to ensure contact of all the acetone with some bleaching powder. The quantity of acetone, however, is too small to be distributed uniformly over the large surface represented by 180 g. of bleaching powder without great trouble and danger of loss of the volatile acetone during the operation. Since, however, the reaction takes place even when the acetone is diluted with water, we are allowed to use sufficient water to suspend the bleaching powder. We should use an amount of water which will give a mixture sufficiently fluid so that, when swirled, we get free mixing. This will be found, in practice, to be about 500 cc. of water. We can add the acetone to this mixture slowly in order to control the frothing.

The reaction is exothermic. The great dilution of the acetone by the water, which has a very high specific heat, will help to prevent much rise in temperature and in actual practice it will not rise much above 50°C. Nevertheless, since chloroform and acetone are so volatile, the question of prevention of loss of these materials by evaporation must be considered. Loss of reactant could be avoided by the use of a reflux condenser and such a procedure could well be undertaken here but it so happens that consideration of the next step of the isolation of the chloroform leads to an alternative. The chloroform, as has been stated, is soluble to the extent of 0.8 per cent at 20°C. in water. It is denser than water. The removal of chloroform by the use of a separatory funnel at the end would involve the loss of 4 g. of chloroform in the 500 cc. of water or an extraction of the water with another solvent. Furthermore, the lime is insoluble in water and may obscure the dividing line between chloroform and water. It is also not unlikely, since the mixture is alkaline, that the chloroform will tend to form an emulsion with the water and be inseparable from it by means of a funnel (emulsions, pp. **43, 67, 110**). The alternative to separation by a funnel would be a distillation of the chloroform. We have not yet considered the problem of distillation of mutually soluble liquids but the problem of the distillation of almost mutually insoluble liquids, such as is presented by chloroform and water, is simple and leads immediately to an answer to the question of isolation of the chloroform.

Distillation of Mutually Insoluble Liquids

The chloroform is almost insoluble in water. The distillation of such a mixture follows a course entirely different from that taken by a mixture of mutually soluble substances. The latter case is discussed under fractional distillation (p. **107**). With immiscible liquids we may assume that the vapor pressure of the one is not influenced by the vapor pressure of the other. In distillation the total vapor pressure is equal to that of the surrounding atmosphere and is made up, therefore, of the partial pressures of each component at the temperature of distillation. It follows that the boiling point of the mixture must be somewhat lower than that of the lower-boiling component. Distillation continues at this temperature until one component is entirely distilled. If P_1 and P_2 are the partial pressures of the components at the boiling point they will appear in the distillate in this molecular proportion and if M_1 and M_2 are the respective molecular weights the distillate will contain M_1P_1 g. of the one to M_2P_2 g. of the other. In the present case the mixture will boil at some temperature below 61°C. (the boiling point of chloroform) and the distillate will obviously contain more chloroform than water (the molecular weight of chloroform being greater than that of water and the partial pressure of the chloroform, whatever it is, being undoubtedly greater than that of water near 61°C.). We should therefore expect to find all the chloroform (20 g.) distilled over in the first 40 g. of distillate. It is true that insofar as chloroform is soluble in water we would expect the distillation to resemble to some extent a fractional distillation wherein, in the case of substances with a boiling point difference of 40° C., a large amount of the lower boiling substance would be retained in the distilling flask after such distillation of a small fraction of the total mixture. However, chloroform is far less soluble in hot water than in cold water so that much less than 4 g. will be left in the distilling flask and the method can be recommended (distillation, pp. **14, 24, 70, 107**).

Preparation of Chloroform

Steam Distillation

Most organic substances are insoluble in water. Many organic compounds, also, have an appreciable vapor pressure near 100°C. Distillation in steam is, therefore, a possible operation on many organic compounds. Since, however, it results in carrying over all material that is volatile in steam it is obviously not a method of purification that can be applied logically to the isolation of a single pure substance unless that substance happens to be the only one present in the distilling flask that is volatile in steam. It is a very useful method, however, for partial purification. It often happens that the wanted material is mixed with inorganic material which prevents the use of a simple extractant for the organic material. Thus it may be mixed with alkalies which result in the formation of an intractable emulsion with all the desirable extractants, or it may be mixed with a bulk of inorganic material which obscures the dividing line between extractant and the rest of the contents of the extraction funnel. In other cases some sticky material, as tar, may prevent thorough contact of the extractant with the mixture; or the mixture may contain colored material which clouds the dividing line of the extractant. Steam distillation can then be thought of as an alternative to extraction. This, indeed, is its most usual role in laboratory work though it is sometimes used for the isolation of organic substances which cannot be distilled at atmospheric pressure because of decomposition. A substance, that is to say, may have a boiling point so high that it cannot be distilled at atmospheric pressure without decomposition and yet be stable at 100°C. and volatile enough to be carried over in a steam distillation. A vacuum distillation (p. 74) is, however, more commonly used in such a case.

The method of steam distillation is convenient for most organic compounds that boil below 300°C. and are insoluble in water. Even when the vapor pressure of the material is as low as 10 mm. at the boiling point of the mixture, the method may be recommended. In such a case, the mixture of compound and water would boil at some point a little below 100°C. where the vapor pressure of the water is 750 mm. and that of the material 10 mm. (making a combined vapor pressure of 760 mm. which we assume to be the atmospheric pressure). The molecular ratio in the distillate (i.e., of the compound to water) will be indeed small (10 to 750), but it is more than likely that the weight ratio will be much larger since most organic compounds of such low volatility would have molecular weights much higher than that of water. The weight ratio would probably be in the region of 1 to 10. Thus a distillate of 1,000 cc. of water would carry over 100 g. of material and a rapid steam distillation would give this result in an hour. The method is convenient for insoluble solids as well as liquids. The exit tubes should be safeguarded against clogging in such case. With a low-melting solid, clogging can usually be avoided by allowing the condenser to get warm on occasion. For high-melting solids it may be necessary to resort to the use of a glass rod or some mechanical contrivance. The end of the steam distillation is shown by the non-appearance of insoluble matter in a test portion of the distillate.

Superheated steam is sometimes used either as a means of decomposing certain compounds into desired fragments (sulphonic acids particularly, which see) or for the steam distillation of less volatile organic compounds.

In the use of superheated steam, the contents of the flask must be kept at some temperature above 100°C. Any water therein must therefore contain some material in solution which can raise the boiling point of the solution to the desired point (inorganic salts, sulphuric acid, etc.) or the work must be done without water in the distilling flask. In performing a distillation with superheated steam it is necessary to have some measure of control of the temperature of the entrant and exit steam. A thermometer for entrant steam is no guide to the temperature of the exit steam or to the temperature of the distilling flask. The difference in temperature between entrant and exit steam can, of course, be controlled by a flame under the distilling flask, but, since the temperature varies considerably from point to point, it is desirable that distillations with superheated steam be carefully described if reproducible results are to ensue. The temperature of the entrant steam and the exit steam, and the dimensions of the distilling flask—or similar data—are items of description which ought to be provided.

Separation by Steam Distillation

In general, steam distillation separates only materials of markedly different vapor pressures at 100°C. If a mixture of substances, all insoluble in water and all having an appreciable vapor pressure at 100°C., were subjected to steam distillation we could not expect any separation of individuals, but we could expect a preponderance of the more volatile individuals in the first portions of distillate. By stopping the distillation at a certain point, or by separating successive portions of distillate, we could achieve at best only a partial separation of individuals. When the mixture consists, however, of substances which vary considerably in solubility in hot water though they have much the same vapor pressure at 100°C. we may achieve much greater separation than before. The insoluble substances would behave as we have already indicated, but the soluble substances would behave more nearly in the manner of the components in a fractional distillation. The insoluble substances would all be driven over while the soluble substances would be largely retained in solution. Similar considerations apply to mixtures of substances which differ in their ability to form compounds with water or with dilute acids or bases. One compound may be driven over in steam while the other is retained in water solution as a compound. Organic compounds having nearly equal vapor pressures at 100°C. would, in all probability, have nearly equal boiling points; therefore steam distillation can sometimes be used for the separation of substances with nearly equal boiling points. From what has been said, however, the separation really depends on a factor of compound formation or solubility and could presumably be achieved by the use of solvents.

Apparatus for Steam Distillation

The apparatus shown for steam distillation includes a steam generator. Where steam is available from a steam line we can dispense with the generator. A flame is usually placed under the distilling flask so that the level of water in the flask can be kept constant: otherwise the con-

Preparation of Chloroform 47

densation of a certain proportion of steam in the flask leads to a gradual increase in the volume of the water. When very wet steam is being used it may be well to use a drier (figure 9) just in front of the entrant tube. With such a drier, indeed, a short steam distillation can be performed without the aid of a flame under the distilling flask. In the pres-

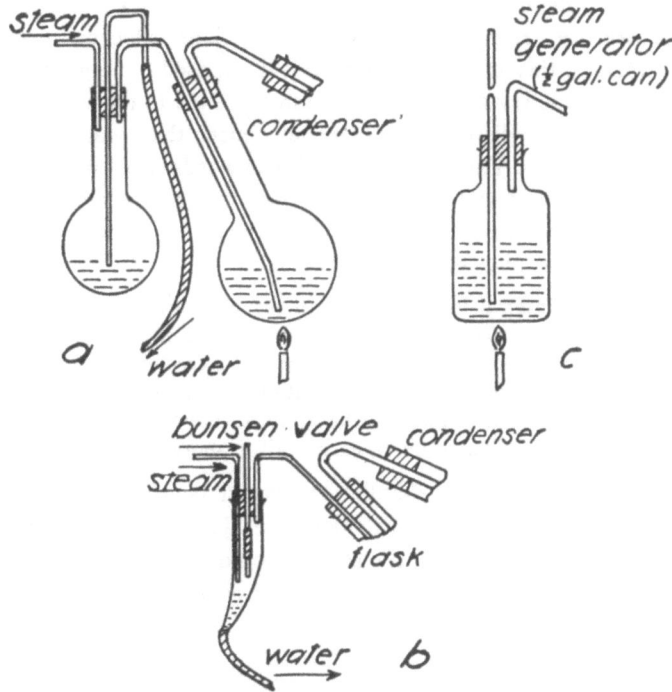

Fig. 9. (a) Steam distillation with a drier; (b) steam distillation with a drier and a Bunsen valve; (c) steam generator.

ent case, for instance, since we need only about 50 cc. of distillate and the distilling flask is ample, we do not need a flame if the drier is used. The drier made from an adaptor is adjustable. Since it cannot fill up, it is obviously preferable to the other drier for long work. There is one objection to the use of the adaptor drier: either very wet steam or fluctuation in steam speed can lead to a backflow of material from the flask into the drier and hence into the waste pipe. The best apparatus is one with automatic safety devices. The drier shown here has a safety valve of simple construction that can withstand gentle pressure inward but is very sensitive to outward pressure. It is merely a piece of rubber tubing slit finely with a sharp knife or razor. It connects an upper tube open to the air and a lower closed tube. The sensitiveness will depend on the length and number of the slits and the thickness of the rubber tubing. It can be tested by cooling the adaptor, after a little steam has been through, while the exit screw clamp to the waste pipe is tightly closed: the contents of the flask should run back only a little distance before the valve operates to let in air. The valve has comparable uses on wash bottles and the like where it is exposed to no great pressure. The bend in the entry tube enables steam to be led to the very bottom of the flask and so to disturb any material denser than water which may have collected there. The flask is sloped backwards so as to minimize the danger of crude material being driven over into the receiver in a froth. A flask with a wide long neck is usually advised for steam distillation so as to minimize creeping effects.

Creeping of a material in solution or in suspension is always possible in a distillation and is the more likely to occur when a stream of vapor moves along at a fast rate. A good demonstration of a creeping effect can be observed by allowing benzene or ether solutions to evaporate in a wide dish. The material in solution is likely to be found to have crept over the edge of the beaker when the benzene or ether has all evaporated.

The condenser in a steam distillation should be efficient, preferably a long Liebig condenser attached to the water jacket by rubber tubing. If an all glass condenser is used, the steam should be carried by a glass tube past the glass joint between inner tube and condenser jacket; the steam should not be driven over very fast.

Isolation of the Chloroform

Distillation of the chloroform in a current of steam is no more than distillation of a mixture of chloroform and water. Since the chloroform is volatile enough to be driven over for the most part in the first 50 grammes of distillate, there seems to be no good reason for maintaining an inlet of steam into the distilling flask in this exercise (with less volatile material it would be advisable to use the usual arrangement of apparatus shown in figure 9a which allows continuous operation for any length of time).

Boiling the reaction mixture so as to collect the first 50 grammes of distillate could be performed by attaching a bent tube to the reaction flask and a condenser as shown in figure 9b. There is, however, danger that local superheating may result in bumping when mixtures containing

Preparation of Chloroform 49

large quantities of insoluble material are boiled; in such an operation it would be best to use a sand bath to make the heating uniform.

As it is best to isolate the chloroform by distillation, and preferably, for the sake of experience, by steam distillation in the usual apparatus in figure 9a, we can now arrange the whole procedure to give the minimum of work and the maximum of yield. The apparatus can be set up for steam distillation with a very large flask as described. Through the stopper a dropping funnel is arranged to lead in the acetone diluted with water. The flask contains the mixture of water and bleaching powder. The acetone-water mixture is run in slowly below the surface of the water with occasional swirling. Reaction should proceed with an increase in temperature and frothing. When frothing starts, the acetone should be withheld and swirling discontinued for a few minutes or until the frothing subsides. When all the acetone has been added, the flask is swirled until frothing subsides; finally a current of steam is led in, slowly at first, and later in good stream. Any chloroform vaporizing up to this point will be collected in the receiver. Before the steam is turned on, care should be taken to see that no solid has clogged up the entry tube (it should be tested by blowing). We can dispense with a reflux condenser during the reaction only because we already know: the temperature rise; the fact that the acetone reacts quickly at low temperatures with bleaching powder; and that the dilution of the acetone is large. In these circumstances, little or no acetone should be lost by evaporation and any chloroform that evaporates is condensed and collected in the receiver. The receiver should be kept cold.

The chloroform is separated from the water in the distillate with a separatory funnel. If the distillate contains some bleaching powder or lime driven over in the distillation, it is permissible to add some dilute hydrochloric acid to dissolve the inorganic material as calcium chloride because chloroform is stable to cold dilute mineral acids. After separation, the chloroform is placed in a flask together with about one-quarter of its volume of anhydrous calcium chloride and is allowed to remain in contact with the calcium chloride, with occasional shaking, for a half-hour or more. This removes water dissolved in, or mixed with, the chloroform. It is then decanted into a distilling flask of the proper type (high side-arm for low-boiling liquids). If a lower layer of saturated calcium chloride solution in water appears at the end of the drying, care should be taken to prevent its introduction into the flask. The chloroform is finally distilled over a small flame protected by a chimney (this is more manageable than the usual steam bath) or from a water bath kept at a temperature around 70°C. There is very little impurity and therefore it is permissible to collect chloroform from 58°C. to 62°C.

Drying Agents

We have dried with calcium chloride just before the final purification. Most organic liquids are dried before distillation, usually by standing for many hours in contact with a hygroscopic inorganic material not soluble in the liquid and not reacting with it when moist or dry. The drying agents vary in efficiency from the almost perfect agents such as sodium and phosphorus pentoxide to the almost useless sodium sulphate which has a large vapor pressure when moistened. Calcium chloride itself is commonly used and is efficient, but it cannot be used with alcohols and acids and it is best avoided when dealing with substances which are capable of change in alkaline conditions (aldehydes, ketones) since it is liable to contain lime: in some rare cases it combines with ethers and nitrogen compounds. Calcium oxide (sometimes carbide) can be used for drying alcohols. Magnesium bromide has been used for drying ether and magnesium alkoxides for drying alcohols. Phosphorus pentoxide is used usually in the desiccator and not in direct contact with material. Potassium hydroxide is often used to dry amines and other basic substances, being useful in removing a molecule of water not otherwise removable from most amines. Potassium carbonate, though basic, can be used for acetone, esters, nitriles, phenylhydrazine. Sodium is used for hydrocarbons and the lower ethers though liable to cause rearrangement on heating with ethers of higher molecular weight. Sodium sulphate is neutral and can safely be used at all times: it is not very efficient. Zinc chloride has been used for drying petroleum, being stable at higher temperatures than calcium chloride. In some cases liquids are dried by means of a stream of dry inert gas at a temperature not likely to involve much loss of material. The desiccator, or vacuum desiccator, is usually reserved for solids, but can be used for liquids if not too volatile. Since liquids have a comparatively small surface for evaporation, the time allowed in the desiccator should be more than that usually allowed for solids.

Exercise 5

Preparation of Ethylene Dibromide

(9, 23, 34, 35, 47, 49)

THE ADDITION of halogen to unsaturated compounds of the olefinic and acetylenic type is an operation frequently undertaken in the laboratory. So general is the reaction and, as a rule, so rapid, that the common qualitative test for these types of unsaturation is the observance of the behavior of a given material with a halogen (usually bromine) in a diluent at room temperatures. With certain exceptions, and in certain prescribed conditions, the reaction even provides a quantitative measurement of unsaturation. Thus the "iodine number" is used for the decription of fats and waxes and other naturally occurring unsaturated substances: a number easily reproducible by the average worker and representing the amount of iodine monochloride taken up in an addition reaction. For the synthetic chemist the reaction can be used to approach acetylenic compounds. Thus maleic acid gives dibrom-maleic acid which can then be treated with alcoholic potash to give acetylene dicarboxylic acid (Perkin and Simonsen, *Jour. Chem. Soc.* **91**:834, etc.). The reaction can be used to synthesize glycols, since the dihalogen compound can be converted into diacetate with silver acetate and then hydrolysed to the glycol.

An example of this addition is provided in the preparation of ethylene dibromide from ethylene and bromine. Ethylene, a gas, can be purchased in cylinders but it can also be prepared quickly and cheaply in the laboratory by the dehydration of alcohol. This method, also, provides experience in the working of glass, the control of a source of volatile material, the employment of safety devices in a train of wash bottles, and many other items of technique which should be generally useful in more advanced work. Setting before ourselves the problem of preparing ethylene dibromide in sample (20 grammes to 100 grammes) by the dehydration of ethyl alcohol to ethylene and the conversion of the latter to ethylene dibromide we proceed, as usual, to consult the literature for a comparison of methods and a survey of the potentialities of each method.

The Literature

Ethylene was first described by Deimann and others (*Chemisches*

Annalen [2; 1795], 195, 310, 430). The gas was obtained by the reaction of ethyl alcohol with sulphuric acid. The reaction was accompanied by much frothing and later Wohler (*Ann.* 91:127) suggested the use of sand to overcome this difficulty. Erlenmeyer and Bunte (*Ann.* 168:64) further improved the preparation by demonstrating that a continuous stream of ethylene could be obtained from the same materials by the slow addition of a mixture of alcohol and sulphuric acid to a mixture of the same sort already heated to about 160°C. As sand was not used in their method, it was necessary to heat carefully to limit frothing of the reaction mixture. The use of a little aluminum sulphate, in the Erlenmeyer and Bunte method, was later shown to catalyze the reaction and lead to more satisfactory reaction speed at lower temperatures (Senderens, *Bull. Soc. Chim.* [4], 9:371. [4], 3:824: *Ann. Chim.* 25:491).

The use of phosphoric acid for the dehydration of ethyl alcohol to ethylene was first mentioned by Pelouze (*Ann. Chim.* 52:37) but it was left to G. S. Newth (*J. Chem. Soc.* 79:915) to demonstrate that it was a good preparative method for the laboratory. No charring took place, the frothing was slight and easily controlled, and a very small amount of phosphoric acid sufficed for the continuous generation of ten to fifteen litres of ethylene per hour over a period of days.

Other dehydrating agents are rarely used in the laboratory. Boric acid can be used to give an ethylene of a high grade of purity (Ebelman, *J. Pr. Chem.* [1], 37:353: Villard, *Ann. Chim.* [7], 10:387 [1897]) and alumina has been used, and is now used, for the preparation of ethylene on the industrial scale. This reagent has also been recommended for the continuous generation of ethylene in the laboratory when large quantities are desired over a long period of time (Gomberg, *J.A.C.S.* 41:1414: Sabatier and Mailhe, *Ann. Chim.* [8], 20:298). This same catalyst can be used, to good advantage, for the preparation of ethyl ether from ethyl alcohol at a lower temperature than is called for in the ethylene preparation.

Ethylene dibromide can be obtained from ethylene merely by leading the ethylene into bromine itself. This was first observed by Balard (M. Balard, *Ann. Chim.* [2] 32:375). The reaction of ethylene with chlorine, however, had been observed even earlier (Deimann, *loc. cit.*) and the product of the reaction between ethylene and chlorine (the "dephlogisticated muriatic acid gas" of Deimann) came to be known as "oil cf the Dutch chemists" (ethylene dichloride). Ethylene itself, giving this characteristic reaction with halogens to form an oily volatile liquid. was described as "the oil building gas." The word "olefine" is reminiscent

Preparation of Ethylene Dibromide

of this property of the first known member of this category of unsaturated hydrocarbons.

For the preparation of ethylene dibromide it is not necessary to take many precautions in purifying the ethylene. When ethylene is generated from alcohol and sulphuric acid it contains a quantity of sulphur dioxide, some carbon dioxide, carbon monoxide, traces of higher unsaturated hydrocarbons. The ethylene generated from alcohol and phosphoric acid may contain some saturated hydrocarbons, traces of higher unsaturated hydrocarbons, but of course no sulphur dioxide. With the first method it is sufficient to use solutions of an alkali for removing sulphur dioxide and some concentrated sulphuric acid to remove the traces of unsaturated hydrocarbons. With the second method, since no sulphur dioxide is evolved, it is sufficient to use only concentrated sulphuric acid. The sulphur dioxide must be removed because it reacts with bromine to give hydrogen bromide and sulphuric acid; the hydrogen bromide is then capable of reacting with ethylene to give ethyl bromide. This would not only reduce the yield of ethylene dibromide but would complicate the matter of purification of the dibromide. The concentrated sulphuric acid is used to retain the traces of higher olefines. These are polymerized or dissolved in the sulphuric acid—ethylene itself is only slightly soluble in cold concentrated sulphuric acid and is not attacked by it. These higher olefines are best removed because they, also, can react with bromine. The concentrated sulphuric acid can also serve to retain any alcohol vapor which may be carried over: alcohol is also capable of reaction with bromine. The ethylene so purified is by no means pure enough for analysis: it is, however, free from the contaminants that could lead to difficulty in the purification of the ethylene dibromide or to appreciable loss of yield. The ethylene dibromide is purified by distillation.

Choice of the Method of Preparation

Most laboratory manuals refer to the use of sulphuric acid and of phosphoric acid as suitable methods. With the sulphuric acid method we can anticipate a little difficulty in the control of the frothing. With the second method there are indications that the introduction of the alcohol into the phosphoric acid may be somewhat troublesome (Fisher, *Laboratory Manual of Organic Chemistry,* Wiley and Sons, 1920). In connection with this, however, we can note that Prideaux (*Chem. News* 113:277) varied the usual procedure of dropping alcohol below the surface of hot phosphoric acid and introduced the alcohol as vapor. He reported no difficulty in keeping a steady generation of gas. The difficulty of introducing a steady flow of cold alcohol into hot phosphoric acid will be discussed later. Since there is little to choose between the methods, the phosphoric acid method will be discussed in detail merely because it does need a little more argument than the first method.

Sulphuric Acid Method

For the preparation of ethylene by this method the literature already cited contains the following instructions. A mixture of one volume of alcohol with four volumes of concentrated sulphuric acid is made up in a flask with a capacity of about twenty times the volume of the liquid. The flask is heated carefully until ethylene is evolved and frothing begins. Then a mixture of two parts of sulphuric acid and one part of alcohol (by weight, understood) is run in through a dropping funnel as required. The temperature of the generating flask is regulated, by means of the flame, so that the evolution of gas is not too rapid. Before heating commences, a little aluminum sulphate (about 2 per cent of the original mixture by weight) can well be added to catalyze the reaction. The ethylene is purified as previously stated. An original mixture of about 100 cc. in a 2,000 cc. flask would be advised for the generation of gas at a rate suitable for laboratory work (one to five bubbles a second from ordinary glass tubing) and the sample of ethylene dibromide could be obtained in an hour or two. The arrangement of wash bottles, and so on, can easily be obtained from the corresponding arguments on the detail of the phosphoric acid method. An extra wash bottle would be needed here for the absorption of sulphur dioxide. The quantity of alkali necessary can be determined from the statement that the percentage of sulphur dioxide in the crude ethylene is about 1 per cent and that two or three times the calculated quantity of alkali is used in fairly dilute solution (5 to 10 per cent) so that no sodium sulphite is liable to be precipitated and clog up the entry tube into the alkali. Since heat is evolved in mixing alcohol and sulphuric acid, the mixing should be done slowly with cooling under the tap.

Phosphoric Acid Method

Only a minimum quantity of phosphoric acid is necessary. There is a slight frothing. A generating flask of 500 cc. capacity, however, is ample to take care of the frothing which occurs even when ethylene is being generated at a fast rate of bubbling (three to four bubbles a second through the usual tubing of internal diameter about 4 millimeters). In a flask of this capacity a quarter-inch layer of phosphoric acid is obtained with about 40 cc. of phosphoric acid (use a round-bottomed flask).

Preparation of Ethylene Dibromide 55

Since the alcohol must be introduced below the phosphoric acid, this would be near a minimum quantity in practice. The frothing increases the surface of phosphoric acid available for reaction and for this reason a fast rate of ethylene generation is obtainable with very little phosphoric acid. Whichever generator is used a certain amount of unchanged alcohol vapor and water is driven over. These would dilute the sulphuric acid greatly if allowed to go forward. A cooling bottle is therefore placed next to the generator and it should be large enough to condense all the alcohol that is likely to be unused (in practice this has been found to be about 30 cc. for each hour of work). A 250 cc. bottle should therefore be ample for cooling when preparing a sample of ethylene dibromide. We could use a condenser leading into a receiver, instead of a cooling bottle. The cooling bottle should be kept cold in ice water or cold water which is replaced when necessary as it warms up: a larger beaker can be used to surround the cooling bottle with cold water. From the cooling bottle a leading tube should carry the gas below the surface of concentrated sulphuric acid in order to wash out traces of higher olefines and remaining alcohol vapor as stated before. Figure 10a shows an arrangement of a cooling bottle and two bottles for receiving the gas below the surface of two liquids as it passes forward.

The amount of sulphuric acid advised is about 20 cc. for sample work: it is used to greatest advantage in a receptacle which gives greatest depth for a given volume. The sulphuric acid also gets warm by solution of the alcohol vapor, water vapor, and higher olefines: it should be kept cool in cold water. From the sulphuric acid bottle the gas is led below the surface of bromine. The reaction to form ethylene dibromide is exothermic and the bromine bottle is also immersed in cold water. Surmounting the bromine with a little cold water helps to reduce the loss of the volatile bromine. The end point of the reaction is shown when the red color of bromine disappears and the ethylene dibromide is colorless or, at most, a pale lemon-yellow color because of the inclusion of some impurities.

It is to be noted that bromine. a red liquid, density 3.1, boiling at 59°C. and slightly soluble in water (4 per cent at 15°C.) gives very bad burns when spilled on the flesh and that its vapors are intensely irritating, even dangerous, to inhale in the smallest quantities. In handling it, gloves should be used and it should be transferred from vessel to vessel only in a fume cupboard or near a draft pipe. Great care should be exercised in opening bromine bottles. We know already that it attacks rubber and cork. It should therefore be placed in its receptacle only just before use and kept cool: in the short time of exposure in these conditions the vapors will not seriously attack the rubber or cork stopper of the receptacle.

Preparation of Ethylene Dibromide

From the bromine bottle a leading tube should carry away any bromine fumes. The leading tube should carry the fumes over the surface of sodium-hydroxide solution or into a draft or, better, over the sodium hydroxide solution before being led into the draft pipes since these are attacked by bromine fumes. The ethylene is generated at a temperature from 210°C. to 250°C. and a thermometer is placed in the phosphoric acid for observation of the temperature. At higher temperatures some charring takes place and the dehydration of the phosphoric acid to pyrophosphoric acid and metaphosphoric acid results in a slower evolution of ethylene. The phosphoric acid used should be the syrupy phosphoric acid of commerce (sp. gr. 1.7: approximating H_3PO_4 in composition). It should be noted that alcohol boils at 80°C.: that ethylene dibromide is a colorless liquid, boiling point 130°C., melting point near 10°C., insoluble in water and soluble in organic solvents; that phosphoric acid is not volatile under these conditions. In noting the physical properties we are drawn to observe that the ethylene dibromide should not be cooled in ice water: it may solidify and so clog up the exit tube and perhaps lead to a blow out of some part of the apparatus. The properties of bromine have previously been described.

Apparatus

We shall need (a) a generator with a thermometer, a leading tube for the ethylene, and an entry tube for the alcohol. Since we need firm, airtight connections, a short-necked, round-bottomed flask with rubber stopper is suggested. Some forms of generator are illustrated below. The generator can be supported on a gauze on the ring stand or a tripod. Since it is to be connected to other pieces of apparatus it should be adequately supported by clamps. Next to the generator, resting in a beaker of water on the bench, we need (b) a cooling bottle having an entry tube and a leading tube. The purpose of the bottle is to condense alcohol and water; the entry tube should, therefore, be cut short inside the cooling bottle so that it will not later dip below the surface of the condensed alcohol and water. The next receptacle is (c) the sulphuric-acid bottle fitted with a leading tube below the sulphuric acid and an exit tube to (d) the bromine bottle similarly equipped. From (d), a leading tube carries any bromine vapor into some arrangement for disposing of the irritant bromine fumes. Since the apparatus is to be all air-tight, rubber stoppers should be used. Some glass tubes will have to be connected and, for the sake of strength in such a large assembly of apparatus, the tubes should be connected with pressure tubing wired on with copper wire as shown in the diagrams. The bottles (c) and (d)

Preparation of Ethylene Dibromide 57

must also be cooled. They can be arranged to rest in beakers of water on the bench, or on gauze on the ring of a ring stand, or in any other way for the sake of firm assemblage. In such apparatus as this, there are involved: the technique of glassworking; the washing of gases; and the arrangement of safety tubes against excessive pressure and against variations of pressure. These matters deserve consideration in themselves.

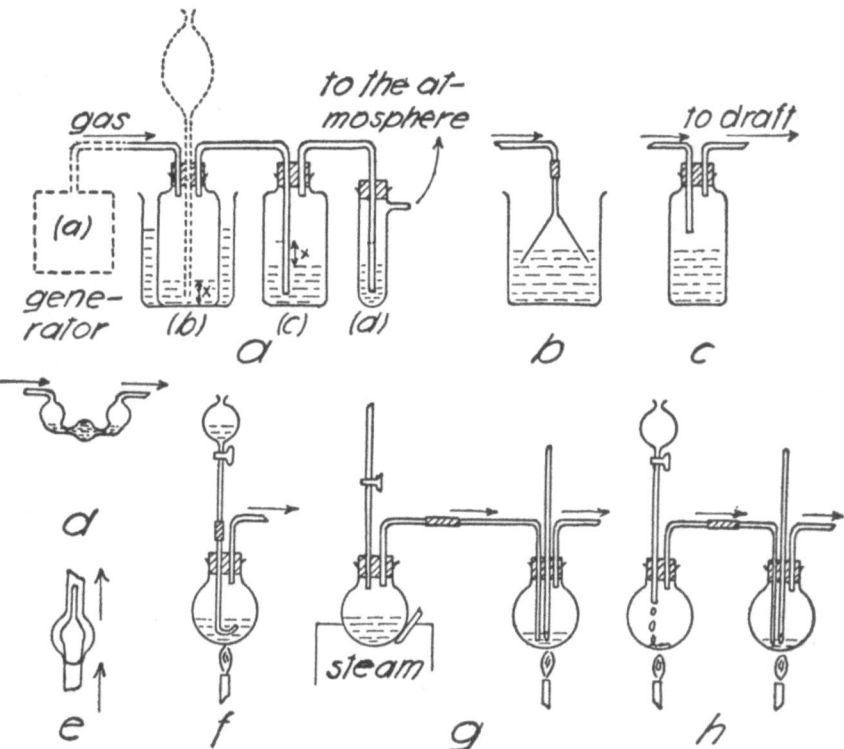

Fig. 10. (a) An assembly of wash bottles; (b) trap for gases; (c) trap for corrosive gases; (d) Pelegot tube; (e) clack valve for air pumps; (f), (g), (h) generators for ethylene.

Safety Tubes

Figure 10a shows a cooling bottle attached to two receptacles containing a liquid through which a gas is passed. This would correspond to the pieces of apparatus (b), (c), and (d) in the assemblage here. The gas comes from a generator on the left. If the gas ultimately comes into the atmosphere at some point on the right, then the outward pressure in the generator is merely that required to drive the gas at the selected rate through the liquids in the receptacles. With ordinary tubing, and with liquids not very viscous, gas can be driven at a fast rate of bubbling by an excess pressure corresponding to about two or three inches of water height (a height of 32 feet of water corresponds to a pressure of about 14 pounds to the square inch). If a tube (the dotted tube in the diagram) is placed in the receptacle (b), extending below the surface of the liquid, it will record the excess pressure above atmospheric in the receptacle by the height to which the liquid rises in the tube. If it is desired for any reason that the pressure in (b) should not rise above a designated limit, it is a very easy matter to arrange a funnel or other receptacle above the (dotted) tube, at a certain height corresponding to the desired limit pressure, which can receive all the liquid driven up from the receptacle. The gas is then free to bubble up through the tube and escape into the atmosphere. As the pressure diminishes, the liquid returns into the bottle and the apparatus is again airtight. Such a safety device against excess pressure will not be needed in this exercise.

The (dotted) tube, however, may be useful as a safeguard against back pressure. If for any reason (condensation, fluctuations of temperature such as are very likely with the ordinary apparatus), the pressure is diminished in the closed system to the left of (b), then the excess pressure in (c) will tend to drive liquid into the bottle (b). At the same time the pressure in (c) diminishes and the pressure in (d) is now in excess of that in (c). As a result the liquid in (d) will tend to flow back into (c); and so on for all similar attachments to the right of (c). The (dotted) tube, however, is open to the atmosphere. The pressure in (b) cannot diminish much below atmospheric before air enters the (dotted) safety tube.

The pressure in (b) can only diminish to the extent of being less than atmospheric by the pressure of the height (x) of liquid. That is, the liquid in the entry tube (c) will rise only a height (x) above the level of the liquid in (c) because the pressure in bottle (c) tending to drive the liquid up the entry tube is atmospheric and will be balanced at the surface of the liquid by the pressure in (b) together with the pressure of the height (x) of liquid in the entry tube in (c). This height (x) will be the same as the depth (x) of the safety tube in the liquid in bottle (b) if the two bottles contain material of the same density.

The safety tube thus prevents backflow from bottle (c) to bottle (b) and moreover prevents any backflow from any similar attachment to the right of (c) into any bottle on its left, because the air from the safety tube is available for all the bottles. The safety tube operates best against backflow when it is adjusted to dip only just below the surface of the liquid.

Figure 10 shows a common safety device used terminally for the absorption of gas in a liquid. It is merely a funnel dipping just below the surface of a liquid in a beaker whose diameter is only slightly larger than the funnel itself. Gases not absorbed escape into the atmosphere. No backflow is possible because the level of liquid in the beaker diminishes rapidly as liquid rises in the funnel and air enters before the liquid reaches the tube above the funnel. This device is suggested for the absorption of bromine fumes that may be driven out of the bottle (c). The bromine fumes could, alternatively, be led over the surface of the absorbing liquid as shown in figure 10c and thence into a draft pipe.

Preparation of Ethylene Dibromide 59

The Pelegot tube may replace the inverted funnel just described or be used in the body of the apparatus. It has, however, a very limited capacity and efficacy for washing. A glass valve is sometimes used as a precaution against backflow of gases when working with pumps.

Though one safety tube is enough in the exercise described here it is sometimes advisable to use more than one safety tube in an assemblage involving many wash bottles. If the gas is passed through liquids of high viscosity in tubes of small bore and if the gas is absorbed very quickly in the absorbing bottle, the frictional effects may be sufficient to prevent quick admixture of air with the gas in the last bottles and may thus cause the material in the absorption bottle to be forced back. That it is rarely necessary to use more than one safety tube can be inferred from the present case where, even with a very rapid reaction between bromine and ethylene, there is no need to use such an extra safety tube. It is worth while, though, to give ample security in all cases by having the entry tubes to wash bottles extend some five or six inches above the level of the liquid.

Washing Gases

The success of the washing process depends on the thoroughness of contact between gas bubble and liquid. For ordinary work, we should choose the conditions which give the greatest amount of contact even when no specific instructions are given. If we are told merely to wash a gas through 20 cc. of sulphuric acid it is assumed that the operator chooses the best conditions with the means available in the laboratory. The longer the contact, obviously, the better the result.

The receptacle for the sulphuric acid should be one so shaped that it gives a maximum depth, and so maximum time of contact between bubble and liquid, for the given volume. Side-arm test tubes, absorption bottles, test tubes, are suggested as having the right shape. The diameter can be anything we choose: it should be noted, though, that fast bubbling from ordinary tubing (4 mm. diameter) would be liable to drive liquid out of an ordinary test tube. The vessel should be at least an inch in diameter.

The size of the bubble, also, has a bearing on the completeness of the washing process. There is more surface in two bubbles each of a certain volume than there is in one bubble of twice that volume. Better washing can therefore be achieved by diminishing the size of the bubble: this can be done by constricting the tip of the tube out of which the gas flows. This constriction will, of course, increase frictional effects so that greater pressure will be required to drive the same volume of gas through the liquid in an equal time. In the exercise at hand it would be, for this reason, unwise to make the tube less than 2 mm. in diameter at its tip.

It is occasionally necessary to wash gases very thoroughly, to scrub them, in fact. For this purpose high towers containing glass beads, pumice, porous tile, or some such material presenting a large surface, are immersed in the washing fluid or merely wetted with it. The gas is led in from the bottom and the bubbles distributed over various routes by the solid material.

Glassworking

Most glass tubing is made of soft glass (approximate composition SiO_2 70% CaO 12% Na_2O 13%) which can be worked with an ordinary Bunsen burner or an air-gas blowtorch. Thick soft glass, such as is often used in bomb tubes, may need an oxy-gas flame for satisfactory work. Hard glass (Pyrex, Jena, and so on) is always worked with an oxy-gas flame because its softening temperature is much higher than that of soft glass. Hard glass can be worked more quickly with a flame, needing less care in heating up and cooling down (annealing), than soft glass; its coefficient of expansion is smaller. Because its coefficient of expansion is small (in Pyrex less than half that of common glasses) hard glass is chosen for work involving heat stresses and is commonly employed for making distilling flasks, combustion tubes, and beakers, for use in the laboratory. Such glass can be quickly heated or cooled but it should be borne in mind that large pieces of apparatus used in the laboratory are all too thick to be heated recklessly even if made of hard glass.

Bending glass tubing. In order to avoid constriction of a tube in bending, the flame should be wide (a fishtail burner or a wing-top attachment to the ordinary burner). The tube should be evenly heated over a long length of tubing (two inches with ordinary tubing; three, four, or five inches with wider tubing). Slow heating in a yellow flame usually gives the best results and the flame should be in the form of an even fishtail. During the heating, the tube is constantly revolved in one direction between the fingers: in any operation in which glass is exposed to a flame, it is to be understood that this rotation is necessary if the object is to be evenly heated. When the glass is so soft that it is difficult to keep the tube straight even by rotating it as fast as possible, it is removed from the flame and bent to the desired shape with a steady movement. The tube, of course, must not be heated to the point of incipient collapse. There should be no constriction in the resulting tube.

Tapering a tube. Heat the tube, with rotation, in a hot wide flame till the tube has collapsed nearly half its diameter. Withdraw from the flame and draw out to the desired diameter at its thinnest part. Break at the required place after marking with a file.

Making capillary tubes. Heat ordinary tubing in a large hot flame, or a small test tube in an oxy-gas torch, until the tubing is collapsed to a diameter of about half the original. Draw out with a pull that steadily increases in speed and power as the glass lengthens. Capillary tubing about 1 mm. in internal diameter can be obtained in lengths from 1' to 3' by this means.

A right-angled joint (T-tube) can be made in the following way. At the desired place of junction in the main tube, a very small hot flame of the blowlamp is played until a circle of red-hot glass can be seen about half the tube in diameter. This area is blown out with the breath to form a small sphere projecting from the side of the tube and about equal to it in diameter (this necessitates closing one end of the tube with a stopper before heating is commenced). The flame is made a little larger and is played on the sphere until it is almost completely collapsed. A strong puff of air into the tube, immediately, blows out the hot portion either into a hole or into a very thin sphere which can be trimmed back into a hole with a piece of metal or wood. The edges of the hole are then trimmed with a small flame so that a projecting shoulder, about a millimeter in length, is left on the sides of the hole. The flame is made larger and the end of the side tube is heated in one part of the flame while the edge of the hole is heated in another part. The end of the side tube should be clean, and is best if recently cut. Ordinary tubing can be cut by marking with a file and snapping between the thumbs held close together. When the hole edge and the end of the tube are at an equal bright-red heat, the glass is removed from the flame and the junction immediately made with pressure and a slight withdrawal after joining. The junction is allowed to cool slightly while the flame is again lowered to a very hot small point. A certain portion of the joint is then heated to red heat in this flame and blown out slightly with the breath. This is done with another portion of the joint and is repeated all around the joint until the shoulder is much larger than before (giving an arc about one-quarter of an inch long from tube to tube). This shaping operation fuses all portions together thoroughly and gives an opportunity for blowing out the thicker portions to a thinner shape. The joint should be well fused in this way and should not be thick at any spot: regularity of appearance is not the first consideration. The joint is allowed to cool in a small yellow flame and to collect a deposit of soot. It is then allowed to cool further in a spot free from draughts of air.

Cutting Glass

Flat glass can be marked best with a diamond or wheel cutter. The glass can then be broken by pressure if the mark is a straight line. If not, a crack must be started at some point by placing the end of a glass rod, hot to redness, at some part of the line. A bright heat may be needed to start the crack. Once started, it can be continued in the proper direction by placing either a very small hot flame or a hot bead of glass, about one-quarter of an inch ahead of the crack. As the crack extends, the glass is moved to a new place: the crack extends towards the hot point of glass. The same method can be used in cutting very wide glass tubing, the starting point or ring being first marked with a diamond or a file. Another method of breaking wide tubing is to revolve the tubing horizontally above a very hot small flame, with the flame tangent to the lower surface. The mark should pass through the center of the little flame in each revolution and the revolutions should be fast. The heating is continued for about ten seconds. By breathing immediately on the tube, a clean break usually results. Many interesting exercises and hints for glasswork are described in *Laboratory Glass Blowing* (Frary, Taylor and Edwards. McGraw Hill. 1928) and in *A Handbook of Laboratory Glass Blowing* (Bolas. Dutton and Co.).

Smoothing glass. Jagged edges of glass can usually be rubbed off by the use of a piece of ordinary heavy copper gauze, using a free heavy stroke. Final smoothing of edges is usually done in a flame. For grinding in glass stoppers emery or carborundum powder, or a very fine washed sand is used. The abrasive is kept moist and the turning done with gentle pressure, alternating the direction of the turning occasionally.

Lubricating glass stoppers. Vaseline is suitable for ordinary work. A mere trace on the side of the stopper is sufficient and in ordinary work a glass stopper which has to be turned at all is kept lubricated (stopcocks; stoppers on separatory funnels; not the glass stoppers of specimen bottles): Shellac is suitable for cementing glass joints. De Khotinsky cement can be applied to warm glass surfaces and sets to a hard gas-tight joint when cool. Ordinary sealing wax is sometimes sufficient. For the lubrication and cementing of glass joints in high vacua the literature should be consulted.

The Gas Generator

Figures 10f, 10g, 10h, show some generators which could be used. The short-necked flask is suggested simply because it gives a tight joint with rubber stoppers on its sharp lip. 10f shows a constricted entrant tube for the alcohol dropped in from above. At a certain rate of dropping (about one drop a second) and with a certain internal diameter at the lip (1.5 to 2 mm.), it is often possible to introduce the alcohol regularly into this generator. It is not dependable, however, because slight changes in pressure at the tip cause the column of liquid to break in the leg of the funnel. When the column has broken it is difficult to start a new flow in a regular manner and, once started, it is liable to break again by a recurrence of the same conditions. If the column breaks, it is best to cool down, by removing the flame, to a temperature near 180°C. and to introduce a batch of alcohol (10 cc. or thereabouts). Then, with the alcohol flow stopped, the acid can be heated to a temperature near 250°C. A flow of ethylene occurs when heating up. The flame is again removed and another batch of alcohol added. Reheating to 250°C. gives another flow of ethylene. In this way the preparation can be completed. The generator is not dependable simply because the smooth working is dependent on the steady vaporization of the alcohol at a certain point in the tube and this is too delicate an operation to control without practice with the particular generator in hand.

Figures 10g and 10h illustrate generators suggested to overcome this difficulty of introducing alcohol: 10g is simply a generator with an outside source of alcohol vapor from a flask which is kept hot on a steam bath. The entry tube into the phosphoric acid need not be constricted although a slight constriction may help to decrease the amount of alcohol that is driven over without reaction. The steam is suggested as the source of heat simply because it is at a steady temperature which we know to be enough to vaporize the alcohol: it will not need attention. The alcohol vapor can be seen bubbling through the phosphoric acid when the generator is in operation and there is little danger of any phosphoric acid running back into the alcohol-vapor generator if the alcohol flask is sunk well down in the steam bath. The (dotted) stopcock is suggested

Preparation of Ethylene Dibromide

as a safety valve, if desired, against the backflow of phosphoric acid. It can be opened when acid is observed to be running back. The generator, we know, is at a pressure slightly above atmospheric: the phosphoric acid will run back, therefore, only to the height representing that pressure if the stopcock be opened. The probable height we have already discussed. The generator 10h is suggested for the maximum efficiency of conversion of alcohol (if that be desired). The alcohol is dropped in at a fixed rate and immediately vaporized above the small flame. A little liquid should always be discernible above the flame. Fluctuations of temperature are more apt to occur with a flame than with a steam bath so that there is more danger of backflow of phosphoric acid. However, a backflow even into the alcohol generator is not a serious matter here since the acid can be poured back into the ethylene generator. The safety stopcock could be used on this generator if desired.

Summary

If a dropping funnel is used in the generator of ethylene, a little alcohol should be run in quickly at the beginning of the experiment so that the leg of the funnel is filled with alcohol. At or near 200°C. the alcohol should be run in at the rate of about one drop a second and the heating continued to the point, between 200°C. and 250°C., where the evolution of ethylene is satisfactory. The phosphoric acid, at this point, will froth a little and the liquid will be filled with small bubbles of ethylene rising to the surface. If the dropping funnel works satisfactorily at a certain temperature it would be unwise to raise the temperature more than a few degrees above that point since the rise in temperature might lead to the breaking of the column of alcohol, with all the attendant bother. When the ethylene dibromide is colorless, or a pale yellow in color, the flame is extinguished and the dibromide transferred to a separatory funnel for washing. It is liable to contain traces of mineral acids (sulphurous acid if the sulphuric-acid method of ethylene generation was used, and hydrobromic acid) and should be washed with cold dilute solution of sodium or potassium hydroxide. Test the aqueous solution after washing to see that it is still alkaline. Then wash once with water to remove any adhering alkaline solution and finally dry with anhydrous calcium chloride. It is usual to use the calcium chloride in generous measure (about one-quarter the volume of the dibromide) so as to present a large surface for reaction with water and so expedite the drying. With occasional shaking the drying should be finished in half an hour or so. The dibromide is finally purified by distillation. The dry bromide can be decanted directly into a distilling flask via a small funnel containing

Preparation of Ethylene Dibromide

a small slug of cotton or glass wool which will retain any calcium chloride accidentally spilled from the receptacle used for the drying. The boiling point of ethylene dibromide is 131°C. but there is very little impurity present and the distillate can be collected from 128°C. to 132°C. The distilling flask should be of the type recommended for the distillation of substances in the medium range of boiling point and the water condenser can safely be used even though the boiling point is close to the temperature (130°C. and above) at which the use of an air condenser is generally recommended for the sake of safety.

Exercise 6

Preparation of Methyl Phenyl Carbinol

(23, 29, 35, 37, 47, 55, 73)

THE PREPARATION of methyl phenyl carbinol from methyl phenyl ketone is an exercise involving the use of metallic sodium and an apparatus for distillation under reduced pressure. Metallic sodium is frequently used by the organic chemist and is a reagent that requires care in handling. A discussion of its use is therefore included, together with a discussion of the manner of distillation under reduced pressure, in this description of the preparation of the carbinol.

The reaction employed is an example of the reduction of a ketone to a secondary alcohol. Reference to the texts recommended for a discussion of group reactions in connection with laboratory methods (see the chapter on the literature of Chemistry) establishes the fact that methyl phenyl ketone is one of a group of ketones for which the use of sodium and alcohol, or sodium amalgam and water, or sodium and moist ether, is usually advised in the reduction to a secondary alcohol. In the case of methyl phenyl ketone itself a further search reveals that the best yields of the methyl phenyl carbinol are obtained when sodium is used in the presence of moist ether, but also that a tolerable yield, about 40 per cent from the ketone, can be obtained by the use of sodium and ethyl alcohol.

Representing the "nascent hydrogen" made available for reduction by (H), the equation for the reaction is $C_6H_5COCH_3 + 2(H) \rightarrow C_6H_5CHOHCH_3$. The hydrogen is represented as coming from the sodium and alcohol (or sodium and water if moist ether is used) by the equation $C_2H_5OH + Na \rightarrow C_2H_5ONa + (H)$ or $H_2O + Na \rightarrow NaOH + (H)$. The phenyl group—$C_6H_5$ is, as far as its chemical properties are concerned in this reaction, equivalent to an aliphatic group such as propyl, butyl, or amyl.

This reaction, however, is not the only one which takes place when a reduction is carried out in this way. A certain amount of a ditertiary glycol is also produced, represented as arising by a bimolecular reduction as follows:
$2 C_6H_5COCH_3 + 2 (H) \rightarrow C_6H_5(CH_3)C(OH)C(OH)(CH_3)(C_6H_5)$.
Such bimolecular reduction products arise even though an excess of sodium is used in these reductions. A description of the preparation of

Preparation of Methyl Phenyl Carbinol

methyl phenyl carbinol from methyl phenyl ketone is therefore a description of (1) the use of sodium to the best advantage, (2) the separation of the products of the reaction from inorganic materials, and (3) the isolation of a pure carbinol from admixture with the ditertiary glycol (a so-called pinacone) and with unchanged material and any other incidental impurities present in small quantity.

The Literature

The reduction has been described in more than one instance in the literature. Neglecting certain minor variations of procedure we are told to "dissolve the ketone in about ten times its weight of absolute alcohol and add pieces of clean sodium at a rate sufficient to keep the mixture vigorously refluxing. Use about three times the theoretical quantity of sodium. When the reaction slows up warm on the water bath and keep the mixture heated until the last traces of sodium have disappeared. When the sodium has all reacted add water and neutralize the mixture with acid. Distill off the ethyl alcohol on the water bath and extract the residue with ether. Remove the ether by distillation and distill the residue under reduced pressure. Collect the carbinol which distills at 118°C. at 40 mms. pressure, at 106°C. under a pressure of 21 mms., and 97°C. under a pressure of 14 mms."

In order to understand the significance of the instructions the information must be supplemented with details which can be gleaned from laboratory tables and specific directions such as are found in most laboratory manuals. With such information these directions can be translated into practice on a specimen of the ketone. A convenient quantity for specimen work would be 20 to 40 g. of the ketone.

Examination of the Procedure

The physical properties of the materials are easily found. Acetophenone is a liquid, density 1.026, soluble in alcohol and ether and in most organic solvents, insoluble in water. Methyl phenyl carbinol is a liquid of similar solubilities, density 1.003, boiling at 202°C.-204°C. with slight decomposition in ordinary practice. The pinacone is a solid of roughly similar solubilities, not volatile below 300°C. and slowly decomposed at or near that temperature. Sodium ethoxide is a colorless solid, very hygroscopic, unstable in moist air and immediately hydrolyzed by water. Sodium is a soft metal which melts at 97.5°C. and boils at 880°C. Its density is 0.97. It tarnishes very rapidly in the air by reaction with atmospheric moisture and reacts explosively with water. With alcohol

Preparation of Methyl Phenyl Carbinol 67

it reacts to furnish hydrogen and sodium ethoxide: this reaction is strongly exothermic but invites no danger of fire.

Further examination shows that the amount of sodium required to keep the suggested amount of alcohol boiling (ten times the weight of the 20 g. sample of ketone) would be a cubic centimeter or so every few minutes. The procedure is now understandable. The quantity of sodium is more than theoretical but is the one found to be desirable in practice. Since the reduction is carried out in a very dilute medium wherein the sodium can react with the solvent alone as well as with the ketone and solvent together, it is not surprising that a large excess of sodium is used. The amount of alcohol still seems excessive because the weight of alcohol required for reaction with the sodium would be about twice the weight of the sodium. This is explained by the properties of sodium ethoxide and sodium. The former is a solid of limited solubility in alcohol which must not be allowed to precipitate on the surface of the sodium and thereby prevent further reaction of the sodium. Since the solution is to be diluted later with water, it would be dangerous to have particles of sodium left in the solution hidden inside a deposit of the ethoxide. The vigorous refluxing is called for as a result of the observation that the yield is better when the reaction is allowed to proceed at the maximum controllable temperature. The mixture is kept hot until all traces of sodium have disappeared in order to keep the reaction going as fast as possible and to prevent deposition of sodium ethoxide which is liable to come out on cooling. The removal of organic products with ether by extraction is a normal procedure when inorganic material and water are present. The removal of alcohol before extraction with ether is also reasonable since alcohol is present in large quantity and is soluble both in ether and in water. It must be removed at some stage or another and it is best removed at this point. When an ether extraction of products is made from a mixture of alcohol and water, it means extraction from a medium in which the products are more soluble than they are in water alone, necessitating the use of larger quantities of ether and unnecessarily increasing the bulk of solvent to be removed by distillation later. The neutralization of the mixture is also desirable since emulsions are liable to form with ether in alkaline solutions. The removal, by distillation, of ether (b.p. 35°C.) and alcohol (b.p. 80°C.) should involve only a little loss of carbinol (b.p. 202-204°C.). It has been stated before that in distilling mixtures whose components have boiling points as widely different as in the case of the present solvents and the carbinol, the distillation should not lead to much loss of the

Preparation of Methyl Phenyl Carbinol

higher-boiling component around the boiling point of the lower-boiling component.

Performing the Experiment

The first part of the procedure is the reduction with sodium and alcohol. There is nothing in the instructions to indicate that the procedure cannot be interrupted at will, but it can be seen that the interruption of the reduction process would involve at least reheating on the steam bath to dissolve any sodium ethoxide which comes out on standing. We should plan to complete the reduction, if possible, before interrupting the experiment.

In general an exercise would not be interrupted except at a point where no further reaction can take place between the materials involved or between the materials and the moisture, or carbon dioxide, or oxygen, of the air. This means that it is usually not advisable to interrupt a reaction or to leave reaction products in contact with one another, but rather to push forward in one working period to the point where purification of the product begins.

The Reduction

Apparatus must be arranged so that sodium can be added to a mixture from which alcohol is evaporating and being refluxed. The simplest way to do this would be to use a flask surmounted by a reflux condenser. Sodium could then be added through the top of the condenser and the only points needing attention would be precautions against blocking the condenser tube by pieces of sodium and dryness of the condenser tube. The reflux condenser should be one in which there is no constriction at the dropping end and the pieces of sodium should be cut so as to slide smoothly through the tube. Another arrangement could be used instead of this: one in which an addition tube, with a water condenser attached to the side arm, surmounts the flask. In this arrangement (figure 11) the sodium would be added through the straight arm at intervals and the straight arm would be closed with a stopper when sodium is not being added. Using a 20 g. sample of ketone the volume of alcohol would be approximately 250 cc. (density of alcohol 0.8) and the flask to be used should be from 750 cc. to 1,000 cc. in capacity, preferably round bottomed since distillation is to be performed later, and attached by rubber stoppers since these are not damaged by boiling alcohol.

Sodium is usually kept under kerosene or naphtha. These liquids, composed of hydrocarbons mostly of the paraffin series, are fractions of petroleum distillates occurring approximately in the range 150°C. to

Preparation of Methyl Phenyl Carbinol

300°C., or 120°C. to 250°C., respectively. Sodium can easily be cut with a knife. It should not be exposed to the atmosphere any longer than necessary in the operation.

One way of ensuring as little exposure as possible would be to weigh the sodium under kerosene in a beaker. The tarnished sodium supplied is first pared with the knife in large pieces and immediately replaced in kerosene so that bright pieces are available. These are transferred, in the correct amount, to the beaker of kerosene. Thereafter the larger pieces can be subdivided into pieces to fit easily into the addition tube. These should be about an inch long so that, on the addition of one or two pieces every few minutes, the mixture can be kept refluxing. Smaller pieces would simply mean more

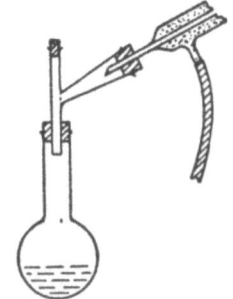

Fig. 11. Addition tube with condenser attachment.

work in addition and more bother in preparation of the sodium. Since the pieces of sodium are wet with kerosene which may later contaminate the distillates (its boiling range includes the boiling point of the carbinol and it is soluble in ether) they should be individually freed of kerosene by touching with filter paper on all surfaces. A few pieces of filter paper should therefore be ready on the desk. Each piece of sodium is placed on a dry portion of filter paper just before addition and touched on all surfaces by turning up the paper and lightly pressing it on the sodium. A further discussion of the use of sodium is found at the end of this chapter.

Toward the end of the reaction the flask should be placed on a steam bath and kept hot until all traces of sodium have disappeared.

Collecting the Reaction Products

The instructions call for the addition of water and neutralization when the reaction is over. The amount of water is not specified and it can be taken to be an amount convenient for dissolving the salt formed on neutralization so that the extraction later is not complicated by the presence of inorganic solids.

Any acid can be used for neutralization but it is usual to use acetic acid or hydrochloric acid rather than nitric or sulphuric acid. Nitric acid is an oxidant and sulphuric acid is liable, when concentrated, to decompose most organic compounds. If acetic acid is used we must anticipate the presence, later, of traces of acetic acid in the ether extract. A trace of acetic acid is usually easily removed in distillation, but the mineral acids, which also are soluble to some extent in ether, may cause decomposition. Hydrochloric acid is, of course, more volatile than the others. Sulphuric acid is particularly undesirable because it can often cause continuous decomposition by a process analogous to that of the continuous dehydration of ethyl alcohol to water and ethylene. The decomposition products, in this case, are likely to be more volatile than the concentrated sulphuric acid; distillation would merely drive the decomposition forward by removing water and the organic products and by leaving the concentrated sulphuric acid behind. Whenever, therefore, an ether extract is made from a medium containing traces of mineral acids, it is usual to wash the ether extract once with a solution of sodium carbonate, or with dilute sodium hydroxide or some other permissible base. If acetic acid is present in traces in the ether extract, it is usually not necessary to wash with a solution of sodium carbonate unless the product sought in the ether extract happens to be a liquid of approximately the same boiling point as acetic acid.

In the present case the instructions would mean the addition of a few hundred cc. of water and an amount of glacial acetic acid (density 1.05, approximately 100 per cent acetic acid) sufficient to convert the sodium to sodium acetate. If the weighing of the sodium is rough and we want as little acetic acid contaminating the extract as possible, it would be proper to test the end point with litmus paper.

Litmus paper is meant to be used in water solutions and not in solutions in organic solvents. To test the end point, therefore, a little of the mixture should be taken out on the end of a glass rod and shaken with water so that the alcohol is diluted greatly. The water solution so obtained is satisfactory for the test. A general instruction for testing mixtures of organic materials for acidity or basicity is: that the test must be made in water solution or, at worst, in a water solution containing only small proportions of organic solvents miscible with water.

Removal of alcohol by distillation is a process usually performed on the steam bath. The difference of 20°C. between the boiling point of alcohol and water is sufficient to overcome the cooling effect of the air on the alcohol vapor and so to drive over most of the alcohol. It is to be noted, however, that such a head of temperature would be insufficient for vessels with large areas exposed to the air. It may be worth while, at this point, to consider more closely the matter of efficiency to be expected in such removal of alcohol on the water bath.

Preparation of Methyl Phenyl Carbinol 71

It will be remembered that, in the case of distillation of mutually insoluble liquids, the more volatile component is completely removed at some temperature a little below its own boiling point. This occurs with the admixture of a little of the component of higher boiling point and it occurs, it is important to note, without any rise of the temperature of distillation. If the solvent to be removed were benzene (b.p. 80°C.; almost insoluble in water) we could safely say that once the benzene started to distil it would be almost completely removed if the heating arrangements were undisturbed. This is not the case when dealing with alcohol even though its boiling point is also about 80°C. Alcohol is completely miscible with water and handbooks show that water-alcohol mixtures boiling at 86°C. have 20 per cent alcohol in the residual liquid and 72 per cent alcohol in the vapor; at 90°C., 10 per cent in the residual liquid and 61 per cent in the vapor; at 95°C., 4 per cent in the residual liquid and 39 per cent in the vapor. This is illustrative of distillations of mutually soluble liquids. The matter will be referred to again but it is easy to see that the removal of all the alcohol will not be accomplished on the water bath because the water bath itself is only a few degrees above the last quoted temperature of 95°C. There is not sufficient head of temperature (5°C.) for a distillation even in small flasks, and we must conclude that the residual mixture will contain between 4 per cent and 10 per cent of alcohol even if a towel be wrapped around the flask and the heating be continued to the point of no further distillation. The removal of alcohol on the water bath should, in view of the argument, be done as efficiently as possible. The flask should be sunk well down in the steam bath, wrapped with a towel, and heated until no more alcohol distills or at least until the distillation is very slow. There is no need to use a distilling flask: the original flask can be fitted with a bent tube leading to a condenser.

Extraction with ether is the next step. It seems curious that the operator should be advised to extract with ether when the carbinol is insoluble in water. At first glance the use of a separating funnel would seem to be the logical operation. It must be remembered, however, that the carbinol is soluble in alcohol and therefore, presumably, soluble to a certain extent in the residual alcohol-water mixture. The solubility in this mixture is small, of course, but the mass of the mixture is large compared with the expected yield of carbinol. Furthermore it is not easy to separate a few grammes of an oil from a large volume of water in satisfactory fashion with a separatory funnel. The transfer from flask to funnel will mean the loss of a cubic centimeter or so of the oil by spreading over the flask surface and a similar loss on the surface of the funnel: perhaps altogether a loss of 4 or 5 cc. in further transfers to distilling flask and so on. The use of ether, in two or three extractions, should prevent much of such loss. The volume to be handled will be increased to such an extent that the loss on surfaces will be a negligible fraction of the whole.

For such reasons as this it is usual to extract small quantities of oils from admixture with water and inorganic materials by ether even when the solubility of the oil is negligible in the water or the mixture being handled. With a volume of 400-500 cc. of the mixture, containing a little alcohol, we should extract at least twice with 100 cc. or 150 cc. portions of ether, the ether being soluble to some extent in water (7 per cent) and much more soluble in the mixture of alcohol and water.

The instructions do not mention drying the ether solution at this stage. It is usual, however, and, from the list of drying agents on p. 50, it will be proper to choose one which does not react with alcohols (anhydrous potassium carbonate, anyhdrous sodium sulphate) and use it for drying. Ten grammes, or more, would be advised for a 150 cc. ether solution. If possible it should stand overnight.

The last instruction concerns the removal of the ether preparatory to vacuum distillation. The expected yield of crude product is about 20 cc. in volume and the ether solution has a volume of about 150 cc. We can evaporate the ether by distillation on the water bath and transfer the carbinol to the vacuum distillation flask. If this is done, however, the flask used for evaporation of the carbinol must finally be washed once or twice with ether and these washings added to the vacuum distilling flask. Otherwise the loss on the flask surface will be an appreciable fraction of our yield. If the transfer is performed in this manner, the ether from the ether-washings must again be evaporated from the vacuum distilling flask. A general instruction can be made here: in all transfers a washing must be done if loss is to be avoided. The amount of loss of material will depend on the size of the container and the extent of dilution of the material in the container. It would be hardly worth while, however, to wash a flask which contained in the beginning only a very dilute solution of the material and this fact makes the following technique for the transfer admissible. The flask to be used for the vacuum distillation is placed on the steam bath and connected with a water condenser. One arm is closed (see diagram of a vacuum distillation) and the other arm is fitted with a dropping funnel. The dropping funnel is connected to the flask either with a cork or a piece of pressure tubing, according to the type of Claisen flask employed in the vacuum distillation, and the dilute ether solution is run in slowly while the flask is kept hot. If the ether is run in at about the rate at which ether is distilled off, there need be no fear of sudden boiling and carrying over of material: the flask can be kept as hot as possible on the steam bath and the distillation is quick and convenient. If the ether is added at a rate which does not cause an accumulation of ether in the flask, and the flask is kept

Preparation of Methyl Phenyl Carbinol

quite hot, this method will be found more convenient than the method of transfer after distillation of ether and subsequent washing. The ether solution is transferred to the dropping funnel as desired, using an ordinary funnel for transfer into the dropping funnel. When all has been added, the dropping funnel is removed, the condenser disconnected, and the Claisen flask set up ready for vacuum distillation.

It is to be noted that ether is never distilled with a flame but only on the steam bath and that the stoppered receiver for the ether should be a side-arm flask which is fitted on the side arm with a rubber tube leading to the floor so as to conduct away any ether vapor to a safe place. Never distill ether with flames located nearer than ten feet: it is dangerously inflammable. Other highly inflammable and very volatile solvents (particularly carbon bisulphide, petroleum ether, ligroine) should also be distilled in this way into a receiver not open to the bench.

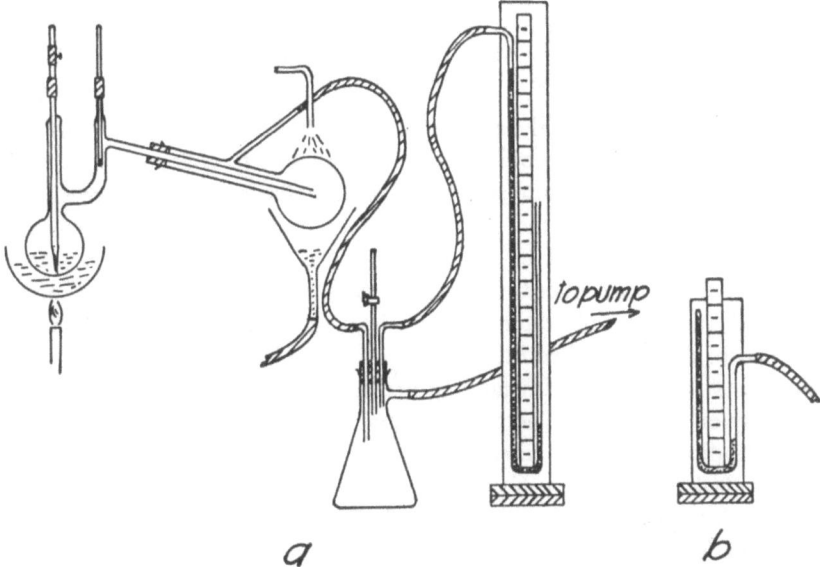

Fig. 12. (a) Apparatus for vacuum distillation; (b) small closed tube manometer.

Purification by Distillation under Reduced Pressure

The apparatus shown in figure 12a can be used for distillation under reduced pressure (distillation in vacuo, vacuum distillation). It is a combination of a water pump, a safety flask with a manometer attached for recording the internal pressure, and a distilling flask with a cooled receiver for the distillate. This form of distilling flask (a Claisen flask) is used in vacuum distillation in order to minimize the danger of material bumping over into the receiver, a phenomenon very apt to occur in vacuum distillations. For the same reason a glass tube drawn out to capillary dimensions is placed in the main neck of the flask so as to introduce into the bottom of the flask a fine stream of air when required (the vapor phase is thus introduced at the place where superheating occurs and tends to prevent a delay in the appearance of that phase and thus prevent superheating). The entry of air is regulated by the screw clip placed above it. The extra piece of glass tubing above the clip is useful as a shoulder for the rubber tube which is subjected to the pinch of the screw clip. If the screw clip is used without such a support the rubber tubing is liable to remain collapsed, under the pressure of the atmosphere, after the screw has been loosened.

The receiver shown provides sufficient cooling surface for materials disstilling above 100°C., but it is not satisfactory for lower boiling distillates. For the latter, the ordinary water condenser could be used, being fitted into the apparatus between the distilling flask and receiver in the usual way.

The receiver itself should be either a round-bottomed or a thick-walled flask, since the ordinary conical flask is very liable to collapse when used in this way. The diagram shows a distilling flask used as a receiver: this can be attached to the pump, or safety flask, by means of the side arm. Between the receiver and the pump, a manometer and a safety flask should be introduced. The safety flask is necessary because, occasionally, variations of water pressure lead to a backflow of water from the pump. A thick-walled filter flask, as shown, is handy for the purpose since it is suitable for withstanding pressure, has a large neck, and rests securely on the bench. The pump is attached to the side arm. The glass tubes leading to the manometer and to the receiver are introduced to the safety flask through a rubber stopper. The figure shows also an extra stop cock tube attached to the safety flask. This is not absolutely necessary but it makes the operation of bringing the pressure inside the apparatus back to normal a simpler matter than otherwise. Any connections in the glass tubing should be made of thick-walled rubber tubing (pressure tubing). The stoppers should be rubber stoppers which are large enough not to be forced in by atmospheric pressure when the pump is working.

Manometers

Two forms of manometer are shown. The simple manometer attached to the safety flask is made by bending glass tubing to the required shape

Preparation of Methyl Phenyl Carbinol 75

and supporting in a stand. A meter stick is supported on the stand between the arms of the manometer. The manometer is filled with mercury by attaching a small funnel to the short arm and pouring the mercury through the funnel. When the pump is working, the mercury rises in the long arm and descends in the short arm; it will come to rest in a position where the difference in height of the mercury in the two arms corresponds to the difference in pressure inside and outside the apparatus. The pump can reduce the pressure to about 10 mm. at best and the pressure of the atmosphere is never much above 770 mm. of mercury. We must therefore arrange to have enough mercury so that the difference in height in the two arms can be 760 mm. while there is still mercury left in the short arm. This can be achieved by pouring enough mercury in the manometer to have a height of about 400 mm. in both arms. The height of the long arm should be, obviously, something in excess of 800 mm.; that of the short arm should be over 500 mm. to ensure a margin of safety against overflow when the mercury returns. The pressure inside the apparatus is given by subtracting the difference in mercury height in the two arms from the height of the barometer at the time of observation. The water pump acts on the simple principle of trapping air inside a fast, small stream of water. Therefore the apparatus will contain water vapor and the internal pressure cannot be reduced below the vapor pressure of the water at the time of using. The vapor pressure of water at 0°C. is the pressure of a height of 4.5 mm. of mercury and at 25°C. is equal to the height of 23.5 mm. of mercury. Consequently the water pump will normally give a pressure as low as 10 mm. on a cold day and on a very hot day perhaps no lower than 25 mm.

The second form of manometer is also commonly used (figure 12b). It is merely a closed bent tube about 100 mm. in height leading to the apparatus at any convenient point such as the safety flask. The closed arm is filled with mercury and the open arm is only partly filled. As the pressure is reduced to the region below 100 mm. the mercury descends in the closed arm. The difference in height of mercury in the two arms gives directly the pressure inside the apparatus since the space above the mercury in the closed arm can be considered to be an almost perfect vacuum. These manometers are filled by first introducing a little mercury into the closed arm, inverting, and boiling this mercury while the other arm is kept under mercury in a beaker or a dish. Air is thereby driven out and, on cooling, mercury enters. If necessary the boiling is repeated. After some usage, these small manometers are likely to contain air in the closed limb and they should be occasionally checked against a standard manometer and refilled when inaccurate. If no standard manometer is available these small manometers must be examined frequently for bubbles at the top of the closed limb. The glass tubing used in making manometers should be clean for the sake of avoiding breakage in the mercury column.

Performing a Distillation

If rubber stoppers are allowable they should be used. All glass connections should be made of pressure tubing. With the material ready in the flask, the screw clip should be turned down so that it is almost closed and the pump turned on to its maximum capacity. The pump should always be used at maximum capacity in order to reduce the chances of backflow into the safety flask when the pressure of water fluctuates, as it is apt to do under the conditions in the usual laboratory. If the material in the flask contains traces of volatile solvents it is advisable to let in a strong current of air for a few minutes while warming the flask slightly. This drives off the last traces of volatile solvents and these are carried away into the water line. If this is not done, the pressure shown when testing out the apparatus will be far above the real capacity of the pump (the vapor pressure of the material being higher than that of water because of the presence of the solvents) and the operator may be deceived into thinking that the pump is bad or that leaks are present in the apparatus. The screw clip is then closed.

If, when using the long manometer, the amount of mercury has been misjudged so that the level descends in the short arm to a point less than a half-inch from the bottom it is best to open both the stopcock at the safety flask and the screw clip, and to let air enter the apparatus until the mercury is again level in both arms. More mercury is then placed in the manometer and the pump brought into play again by closing the stopcock and screw clip. The manometer should show a pressure inside the apparatus of a magnitude within 20 mm. of that expected normally on such a day. When the pressure is unsatisfactory, the apparatus must be tested for leaks. The most probable point of leakage is a cork stopper if one is present. It should have been prepared according to the instructions previously given (p. 25) and it must be firmly forced into place without endangering the hand. The hands are always endangered if glass apparatus is squeezed between them, but if only the fingers and thumbs are used, not the palm of the hands, and if the opposing fingers are not more than an inch apart, the danger of a serious cut is negligible. Any cork stopper will probably need collodion as a covering in order to make it airtight. Shellac or a rubber cement may be used in the place of collodion. If the pressure is still much too high,

Preparation of Methyl Phenyl Carbinol 77

the pump itself should be tested when directly attached to the manometer. Sometimes the seating of a rubber stopper is at fault and sometimes a flask has a minute pinhole present in a badly sealed joint. The junctions of side arms should therefore be examined by painting with collodion or by observation of any changes in pressure occurring when the thumb is firmly pressed against selected areas at the junctions of glass and glass. It is rarely the case that a junction with rubber is at fault but it is usual to hold the joints made of pressure tubing firmly in place by means of copper wire bound around the pressure tubing with pliers. Except on cork stoppers, collodion should not be used unless absolutely necessary.

The apparatus having been tested, the water is turned on at the condenser and the pump and the screw clip adjusted to let in a slow stream of small bubbles of air. The distilling flask is then heated until distillation begins, using a rotating yellow flame if there is no danger of decomposition. The most difficult part of a vacuum distillation is in this raising of the temperature to the boiling point. This is the time when bumping is most apt to occur. The flame should not play on the flask continuously but with intervals of a second or two between each heating with a rotating yellow flame for five or ten seconds. The stream of air should be as large as possible while still allowing a tolerable vacuum of 30 to 50 mm. When distillation has once started, or when a refluxing edge of distillate can be seen rising in the side neck of the flask, it is usually safe to cut off the air inlet almost entirely and to proceed to note temperatures of distillation and amounts of distillate at selected temperature intervals. The same care should be observed when reheating the flask.

In order to collect a fraction or to end a distillation, the flame is removed and the flask allowed to cool for a few minutes. Then the stopcock is carefully opened at the same time as the air inlet, taking care to see that the mercury descends slowly and that the capillary tube is not filled with liquid.

The exercise here undertaken should yield about 8 grammes of a distillate that is almost pure carbinol. It is to be noted that the presence of unchanged ketone in the distillate is not anywhere mentioned as a possibility though it would obviously be in our distillate if the reduction were incomplete (b.p. carbinol, 202°-204°C.; b.p. ketone, 202°C.). If the pressure during distillation is not exactly at one of the quoted points in the directions it can be estimated by taking a rough proportion from the figures given.

At the working pressures of the usual water pump a millimeter difference of pressure corresponds roughly to one degree difference of boiling point. This can be seen to hold roughly for the carbinol and it is a general rule to be remembered. A further discussion of vacuum distillation is given on p. 153.

It may be found that a certain amount of distillate will come over far below the expected boiling point of the carbinol (water and alcohol not completely removed in the collection of the carbinol). If so, the first distillate must be removed when the distilling thermometer shows a temperature near the expected one. The temperature of the distillation thermometer will not rise more than a degree or two during distillation even though the heat is increased as the distillation proceeds. This indicates that there is little or no impurity volatile near the boiling point of the carbinol and a redistillation will be unnecessary. When no more distillate comes over, even though the heat has been definitely increased, the distillation is stopped. It may be advisable to use an oil bath in this first exercise so that no decomposition of the residue occurs by careless heating. It will be seen, in this case, that the carbinol will begin to distill when the oil bath is about twenty degrees higher than the expected boiling point of the carbinol and that the distillation thermometer remains constant while the bath is raised to a temperature about eighty degrees higher than the boiling point. At the same time the rate of distillation will diminish practically to nothing. The oil bath should not be heated over 250°C.: it is unpleasant and there is some danger of fire. Record the yield and boiling point and pressure. Thus: Yield 6 g. b.p.$_{20}$ 149°C.

Sodium

The sodium used in this exercise was not entirely free of kerosene, and, since the kerosene boiling range includes that of the carbinol, we can assume the product will be slightly contaminated with kerosene on distillation. This is a matter usually neglected. If it is taken into consideration, then the petroleum distillates called petroleum ether (b.p. 40°C. to 60°C.) or ligroine (b.p. 80°C. to 120°C.) may be advised for the taring beaker. The original kerosene or naphtha covering the sodium is, of course, already dried by the sodium and, if used in the taring beaker, need not be dried beforehand. The petroleum ether or ligroine, however, may have to be dried beforehand when we are dealing with a reaction in which traces of sodium hydroxide on the sodium would be harmful. The Bouveault reduction of esters with sodium and alcohol, for example, must be sometimes performed with great care to avoid the presence of sodium hydroxide in the materials employed.

Sodium wire is often used. It is made in the sodium press. The clean bright sodium is placed in the dry press and squeezed through by revolving the head screw. As the sodium emerges, in wire form, it is led directly into the receptacle containing the protecting liquid or reaction mixture, exposing

Preparation of Methyl Phenyl Carbinol

for the shortest possible time. The charge of sodium in the press should be, if possible, the whole of the sodium required for the exercise. If the press must be recharged it is best to clean it first by destroying the residual sodium in the press with alcohol and wiping thoroughly dry with a cloth, not neglecting the pin hole of the press. Otherwise the second charge may be contaminated with sodium hydroxide and a little sparking may even occur as the sodium is driven through. The sparking is usually of no consequence but it should not be neglected when dealing with highly inflammable liquids such as ether or petroleum ether.

"Bird-shot" sodium is often used. These small pellets are obtained in the following way. The required amount of sodium is placed in a round-bottomed short-necked flask and covered with dry xylene. The flask is then heated with a flame, under reflux since xylene is inflammable, until the sodium is seen to melt. The xylene should not be brought to a higher temperature than that of incipient boiling. Sodium melts at 97.5°C. and ordinary xylene, a mixture of isomeric dimethyl benzenes, begins to boil near 135°C. The flask is now detached, surrounded with a cloth, stoppered firmly, and shaken vigorously up and down a few times so that the sodium is broken into pellets from 1 mm. to 2 mm. in diameter. The flask is then gently placed on a cork ("suberite") ring or on the ring stand and not disturbed for a few minutes. If the globules coalesce on standing, the temperature was a little too high when the flask was placed on the bench and the operation should be repeated. Xylene is advised simply because it is a hydrocarbon of the right boiling point for the purpose. The round-bottomed flask is the type best suited to withstand the outward pressure which occurs when the flask is shaken, the air inside being heated by that operation. The stopper should be held in place during the shaking by a firm pressure with the hand: the cloth protects the hand against burns. When the xylene is cool it is decanted off and the residual sodium washed three or four times with dry ether by decantation so as to remove the adhering xylene. The sodium is usually required immersed in ether.

Sodium is sometimes required as an amalgam with mercury. The amalgam usually called for is one containing from 1-10 per cent of sodium. A 1 per cent amalgam is a liquid. The higher percentage amalgams are solids at room temperatures and can be stored in well-stoppered bottles without any liquid protection. In making the amalgams it is necessary to avoid breathing the poisonous vapors of mercury and to protect the eyes. If a hood is available the operation is simple. The mercury is placed in a porcelain casserole, or a porcelain pestle, or a hard glass beaker. The sodium is then weighed out under a protecting liquid in pieces about the size of a pea. The pieces are introduced one by one to the bottom of the mercury by spearing on the end of a tapered glass rod. Each piece must be held firmly at the bottom of the mercury until the boiling, or flashing is ended. It may be desirable to protect the hands with a glove but goggles need not be worn if the glass screen of the hood is held below the level of the eye. Before introduction into the mercury the sodium should be freed of adhering oil with filter paper. In making the richer amalgams the sodium must be added, towards the end, quickly enough to keep the mercury quite hot so as to prevent premature solidification.

For comparative purposes it should be stated that the making of a 1 per cent amalgam, without hurry, usually brings the mercury to some temperature above 200°C. If necessary, of course, the amalgam may be kept hot with a flame. A variation of this procedure, using toluene above the mercury, is useful when a hood is not available. Extra care must be taken, in this case, to see that no sodium comes to the surface and fires the toluene. During the addition the heat of the reaction is dissipated in the vaporization of the toluene and the temperature is thereby prevented from rising much above the boiling point of toluene (110°C.). At the end, the toluene is decanted off and the residue is made uniform by heating with a free flame until liquid. At this stage the residual toluene may burn but the flame can easily be extinguished with a cloth.

A general procedure for making amalgams is described by Vanstone (*Chem. News* 103:181). The mercury is added to the sodium in this case. A large piece of sodium is pared and dipped quickly in ether containing a trace of alcohol. The surface becomes bright. The speared lump of sodium is then moved above in the air, for a few seconds only, to allow the ether to evaporate and is then placed under molten paraffin wax contained in a crucible, the temperature being 100°C. All the sodium being under the wax, mercury is added, very carefully at first, to the desired amount. If the amalgam is not molten at the end, it is melted by heating and then allowed to cool under the wax. When the wax is solid it can be easily removed and the amalgam stored in a dry bottle closed with a rubber stopper. The amalgam is often required in small pieces in which case it is best poured, while still warm, on to asbestos paper, metal plate, or even ordinary hard paper turned up at the edges. The efficiency of an amalgam is often gravely affected by impurities in the mercury and a failure in the use of an amalgam can often be laid to the employment of a low grade of mercury. It should be pure and re-distilled.

Exercise 7

Preparation of Ethyl Benzene
(2, 26, 35, 79)

THIS EXERCISE is suggested as an introduction to the practice of fractional distillation and to the use of aluminum chloride. The literature on the preparation deals for the most part with large-scale work and it should be interesting to see what results are obtained when the original quantities are greatly reduced. The reaction employed is a very common one (the Friedel-Craft reaction) and one of the reagents has already been prepared in Exercise 2.

The Literature

Ethyl benzene is usually prepared by the Fittig reaction, using brombenzene and ethyl bromide in the presence of sodium $C_6H_5Br + C_2H_5Br + 2Na \rightarrow C_6H_5C_2H_5 + 2NaBr$. It can be prepared, also, by the Friedel-Craft reaction using benzene and ethyl bromide in the presence of aluminum chloride $C_6H_6 + C_2H_5Br \rightarrow C_6H_5C_2H_5 + HBr$. The aluminum chloride is best used in something approaching molecular proportion although the formulation of the reaction given above indicates that it is a catalyst. The reaction is quite general for the introduction of alkyl groups into aromatic hydrocarbons and the alkyl halide usually employed is the chloride. Ethyl chloride, however, is a gas at room temperatures and is not readily available in most laboratories.

The conditions for the reaction between ethyl bromide and benzene in the presence of aluminum chloride have been described by Behal and Choay (*Bull. Soc. Chim.* [3; 1894], 2: 207). They claimed that a very good yield could be obtained if the aluminum chloride was added in small portions to a mixture of the benzene and ethyl bromide kept warm on the water bath. With 5,000 g. of benzene they used 500 g. of ethyl bromide and added aluminum chloride (something over 100 g.) in portions of 10 g. at a time until the theoretical amount of hydrogen bromide had been evolved. On first addition of aluminum chloride the reaction was warmed to start the reaction and each subsequent addition was made after the reaction induced by the previous portion of aluminum chloride had moderated. At the end the material was thrown into ice water containing some hydrochloric acid and the upper benzene layer which appeared was separated, dried with calcium chloride, and submitted to fractional distillation in a Le Bel-Henninger distilling column of five bulbs. The reaction had been described previously, using lower temperatures and longer time, but the yields were low. Behal and Choay ascribed their better yields to the method of gradual addition of aluminum chloride and the higher temperature of reaction.

Procedure

Ice water is used to moderate the reaction of the aluminum chloride with water. The purpose of the hydrochloric acid is merely that of retaining the hydrated aluminum chloride in solution in the water. Aluminum chloride itself hydrolyzes in water to give some aluminum hydroxide which is insoluble; it also tends to cause an emulsion between the benzene and water layers: hydrochloric acid represses the hydrolysis.

The drying with calcium chloride is done in the ordinary way using sufficient calcium chloride to prevent the formation of any solution of calcium chloride in water from the moist benzene. If any such solution of calcium chloride is seen after standing, it means that insufficient calcium chloride has been used to take up all the water as a solid hydrate and more anhydrous calcium chloride must be added.

The fractional distillation is a distillation in which the material is collected in fractions as the distillation proceeds. These fractions are then separately distilled, in a logical fashion, for the purpose of obtaining a maximum yield of product. Such a distillation, described below, becomes necessary when the material wanted is mixed with other liquids impossible to remove by one distillation. The difficulty of separation of two liquids by distillation increases as their boiling points approach each other. In the present exercise the problem is one of separating the ethyl benzene (b.p. 135°C.) from a much larger quantity of benzene (b.p. 80°C.) and higher-boiling by-products of indeterminate composition which may be present in amount approaching that of ethyl benzene itself. The large proportion of unwanted liquids is also a factor in increasing the difficulty of separation of ethyl benzene in anything approaching totality and purity.

The Exercise

Arrangements must be made for dealing with specimen quantities which would be only a fraction of the amounts dealt with in the literature. We could proceed to cut down all the quantities in the proper proportion. For one-tenth of the given quantities this would mean the use of 500 g. of benzene, 50 g. of ethyl bromide, 10 g. to 15 g. of aluminum chloride added in 1 g. portions. Such an arbitrary procedure is not really rational since it neglects the consideration of factors of time, temperature, and

Preparation of Ethyl Benzene

the like, which may be of major importance and yet unwittingly altered by this simple rule. In view of the fact, however, that the original description in the literature is as meager as it is, this simple rule must suffice for the moment (see Part III for general discussion). We must not be surprised if the percentage yield is altered by the change.

Arrangements must be made for testing the end point. If the original procedure is followed in its entirety a receptacle must be provided for the hydrogen bromide evolved. This may be led into a solution of sodium hydroxide containing a weighed amount of the hydroxide, and the end point will be reached when the solution becomes neutral or just acid. The amount of hydroxide, or alternatively carbonate or bicarbonate, would be that calculated from the hydrogen bromide. It must be realized, however, that ethyl bromide is very volatile (b.p. 38°C.) and will be carried over by the evolved acid to some extent unless the cooling is very efficient; that the hydrogen bromide will be retained to some extent in the mixture; and that the reaction is unlikely to proceed to absolute completion. With this in mind it may be well to use something less than the calculated amount of sodium hydroxide.

If it is decided to use a fixed quantity of aluminum chloride—some definite weight between 10 g. and 15 g. seems proper—then the end point need not be tested accurately. It will be sufficient to note the time intervals at which the portions of aluminum chloride are added and to note the point at which hydrogen bromide ceases to pass through the exit tube. This exit tube, ordinarily, would lead into the draft pipes. By breathing on the exit tube a fume is observed, from the moisture of the breath, when hydrogen bromide is being evolved. It may be proper to protect the draft pipes from the corrosive effects of the acid. In this case the hydrogen bromide is led into water or sodium hydroxide and the testing of the end point is done by breaking the connections of the apparatus at some point in front of the trap and testing with the breath.

If hydrogen bromide is led into water or sodium hydroxide there is always danger of backflow and some arrangement must be made to prevent it (see figure 10).

Powdered aluminum chloride cannot be added satisfactorily through the top of the condenser: it clings too readily to the glass. It must therefore either be added while the condenser is temporarily detached or an addition tube must be used (see figure 11). If a two-necked flask is available so much the better. Since aluminum chloride takes up moisture from the air it must be exposed as little as possible. After powdering quickly it should be weighed into a stoppered flask or bottle.

After collection and drying of the product, the fractional distillation must be undertaken and the product labelled with a description of weight and boiling point. Again it must be noted that a boiling range on a label means the total range of the observed boiling point when the material in the bottle has been submitted to distillation apart from other fractions. It is labelled thus: "ethyl benzene b.p. 135-7°C. 8 g." For the description of a fraction of a body of material the temporary label should read thus: "ethyl benzene, frac. 135-7°C. 8 g."

As the yield is unpredictable, a very careful note should be made of times, amounts, apparatus, etc., in order to compare results with neighbors and suggest improvements in procedure.

Fractional Distillation

When evaporation takes place from a mixture of two liquids the vapor is found to be richer than the residual liquid in the more volatile component. This is true in all cases of liquid mixtures except when an azeotropic mixture (mixture of minimum or maximum boiling point) exists in the case of the liquids under consideration and when at the same time the composition of the liquid happens to be exactly that of the azeotropic mixture.

On distillation, therefore, of any mixture of liquids except that noted above, the more volatile component tends to concentrate in the first portions of the distillate and the less volatile in the later. When the mixture is one of nonmiscible liquids the behavior on distillation is that already described: the distillation proceeds at a constant temperature somewhere below the boiling point of either constituent and continues at that temperature until one component is completely driven over. Thereafter the distillation thermometer records the boiling point of the remaining component. The theoretical treatment is simple and has already been described.

When the mixture is one of partially miscible or completely miscible liquids, however, the theoretical treatment is complicated by the necessity for the consideration of such factors as association and chemical relationship, though the practice of distillation depends on the simple fact of concentration of the more volatile portions in the first part of the distillate (see also Part III).

In the distillation of miscible liquids, the distillation thermometer shows a continuous rise from the boiling point of the lower-boiling component to that of the higher, but the amount of distillate collected in designated ranges of temperature will vary with the proportion of constituents. The greater the difference in boiling points, the smaller the proportion of liquid distilling in the intermediate ranges, and, when the boiling points differ as much as 100°C. or more, it is usually easy to arrange in one distillation for an almost complete separation of the two components with no appreciable loss.

When the boiling points of the miscible liquids are closer together the separation is more difficult. With a temperature difference of 25°C. to 30°C. it is not a simple matter to separate the pure components in good yield, and it becomes very difficult to avoid large losses when separating a mixture of two components whose boiling points are within 10°C. of each other. The

Preparation of Ethyl Benzene 85

method of separation in such cases is merely a logical extension of the principle of distillation and is described below as fractional distillation.

Suppose the first portions of distillate, richer in the low-boiling constituent, to be again distilled. From these first portions, during redistillation, a certain small fraction is obtained which is very rich in the low-boiling constituent and this is laid aside. From each intermediate portion of distillate, on redistillation, a certain amount of material is obtained, in the first runnings, which is of a fairly high concentration in the more volatile constituent. If these first runnings are then combined and redistilled they give a certain small fraction of approximately the same composition as the fraction previously laid aside, and this can be added to the reserved portion for further treatment. A distillation carried out in this way is called a fractional distillation. It is nothing more than condensation and redistillation performed in a logical way to further the tendency of the mixture to separation when it evaporates. Such condensation and redistillation actually occurs during a distillation performed in an ordinary distilling flask but it occurs in a haphazard way on the inside surface of the flask only. The ordinary distilling flask is not well constructed for the purpose of distillation, in fact, and it is often replaced by special flasks and columns in whose construction some effort is made to provide for a maximum of condensation and re-distillation in the interior of the apparatus. These fractionating flasks or fractionating columns often achieve results in one distillation which are comparable to the results of a tedious fractionation by hand with the ordinary distilling flasks and the term fractional distillation is often used, in consequence, to describe either fractionation proper or the use of a fractionating flask or column (distillation, pp. **14, 24, 70, 88,** 107).

Collection of Fractions

The usual method of fractionation is that of collection in defined temperature ranges. When the boiling point of the wanted substance is known, the temperature intervals are chosen with reference to that boiling point and are so arranged that the intervals are smaller near the boiling point than they are further away. In the present case, seeking ethyl benzene of boiling point 135°C., the chosen intervals could well be 80°C.-110°C., 110°C.-130°C., 130°C.-135°C., 135°C.-140°C., 140°C.-160°C. for a five-fraction distillation. For a three-fraction distillation the ranges could conveniently be 80°C.-120°C., 120°C.-140°C., 140°C.-170°C. This method is, of course, purely arbitrary—the temperature ranges are often decided after, rather than before, a preliminary distillation, in order to avoid great differences in bulk in the various fractions. After the fractionation has been carried out for some time in these ranges, the bulk of the wanted material becomes concentrated near the boiling point of the material and it is then proper to change the temperature intervals, shortening the intervals near the boiling point and shortening the whole range of distillation after rejecting the small amounts of material coming over in the extreme ranges. In the present instance, for a three-fraction distillation, the temperatures would perhaps be 110°C.-132°C., 132°C.-138°C., 138°C.-150°C. Ultimately the fractions at 134°-136° would be sought in maximum yield and the product labelled with its complete range when distilled alone.

When attempting to isolate from a mixture some pure material or materials of unknown boiling point, the fractionation is carried out as described above in certain defined temperature ranges. It is of great advantage in this

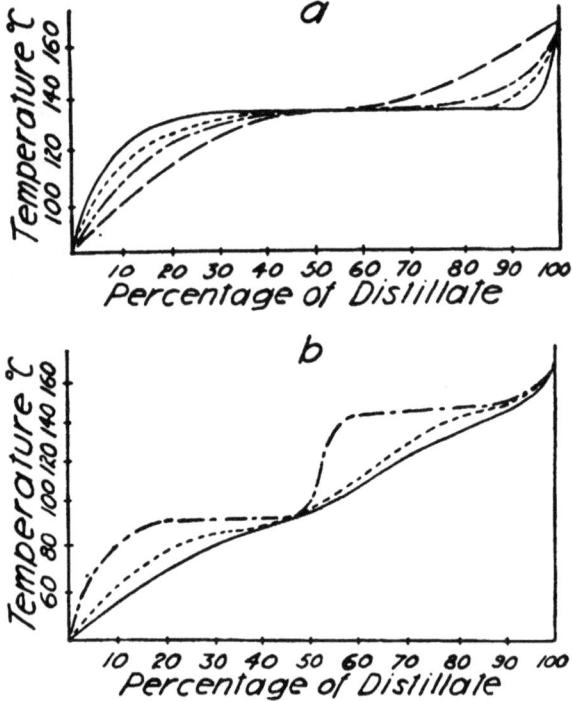

Fig. 13. (a) Fractional distillation of crude ethylbenzene. The abscissae show the percentage of distillate after neglecting losses and discards. (b) Fractional distillation of a substance composed of two individuals boiling at 80°C. and 150°C. The abscissae show the percentage of distillate after neglecting losses and discards.

case, however, to make a note of all the amounts collected and to compare them graphically as the fractionation proceeds. The percentage weight of distillate received up to a designated temperature is plotted against that temperature on the first distillation and the curve so obtained is compared with the curve obtained after one fractional distillation or more. The first curve would more nearly approximate a straight line than the second or the third curve. The curves, indeed, would sooner or later approximate a series of steps equal in number to the number of materials present. Such a change

Preparation of Ethyl Benzene

in the appearance of the curve arises in consequence of the concentration of the bulk of each component near its boiling point, and is illustrated in figure 13a for a crude containing chiefly one substance. In this figure it is evident, even as early as Distillation II, that a large proportion of the material distills near 135°C. In the case of ethyl benzene this is anticipated but even in the case of distillation of mixtures of unknown composition the appearance of such steps would suggest the number of approximate boiling points of the constituents. This information would help in the arrangement of subsequent temperature ranges for fractionation. Figure 13b shows a fractional distillation of a mixture having two chief components.

The Fractionation in Practice

Having decided on the temperature intervals and the number of fractions to be handled, the material is distilled into separate receivers for each temperature interval. The first fraction is then poured into the empty distilling flask and this is redistilled. The material coming over in the first temperature interval is collected in the first receiver and the distillation is then stopped. The second fraction is then added to the residue in the flask and the distillation is started again. Any material coming over in the first temperature interval is added to the first receiver and the distillation is continued into the second receiver until that interval is covered. The distillation is again stopped, the third fraction added to the residue in the flask, and the distillation continued. As before, any material coming over in the first temperature interval is added to the first receiver, in the second to the second receiver, and in the third to the third receiver, and so on. In this way, the usual way, the number of fractions being handled is kept constant and the results are easily put in tabular form as shown below. The first two distillations in the column correspond to the operations described above; it will be noticed that the total weight of material is shown as diminished by losses in operation and by the rejection of material falling outside the total range of temperature chosen. The figures are hypothetical for the distillation of ethyl benzene in the five temperature ranges suggested above.

FRACTIONAL DISTILLATION OF 100 g. CRUDE ETHYL BENZENE

Fraction	Temp. °C.	Dist. I	Dist. II	Dist. III	Dist. IV
A	85–110	10	5	4	3
B	110–130	15	15	10	6
C	130–135	20	25	30	30
D	135–140	22	25	27	30
E	140–170	23	10	4	3
Total loss		10	20	25	28
		100 g.	100 g.	100 g.	100 g.

These results, in graphical form, are illustrated in figure 13a. The amounts of distillate are given as percentages of the total distillate in each case simply for the sake of keeping the ordinates as well as the abscissae of the same total length for each distillation.

At this stage represented by Distillation IV it would be advisable to change to fractionation in new temperature intervals with 110°C. and 140°C.

as the extreme points, making intervals of two or three degrees near 135°C., and rejecting anything outside these new extreme points.

Fractionating Flasks and Columns

Some of the more common types are illustrated in figure 14 below. The simple Ladenburg flask has been previously described (figure 6b) and is often convenient though not very efficient. The Hempel column is usually filled with large heads or small pieces of glass tubing. It is easy to see, in most of the columns, the elements of construction which are introduced on the principle of leading condensed liquid back into contact with hot ascending vapors as often and as thoroughly as possible. The Glinsky column is unusual in having moving parts in the shape of hollow glass balls which are joggled up and down as the distillation proceeds. They provide good contact between condensed liquid and vapor but lead to alterations of pressure which

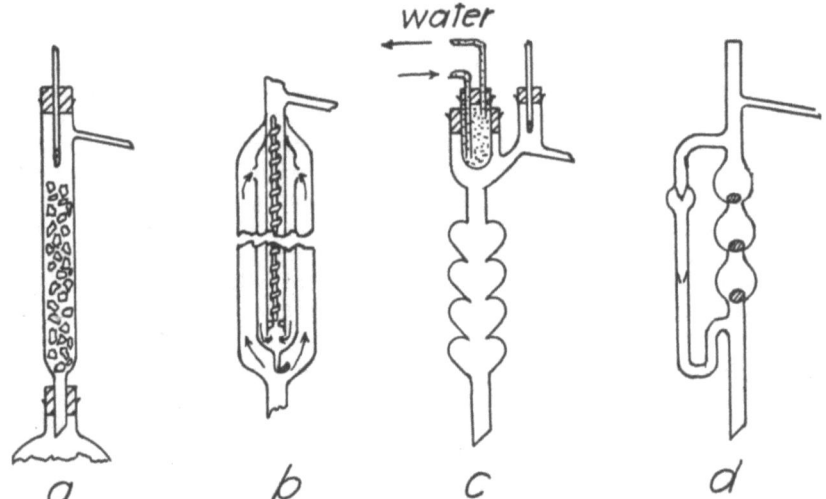

Fig. 14. (a) Hempel column filled with beads or fragments of glass tubing; (b) Widmer column: the arrows show the passage of the vapors on a return path; (c) Young pear column with an attached cooling point for the observation of a reflux ratio; (d) Glinsky column with moving hollow beads.

tend to make the distillation unsteady: the side tube is introduced for the correction of this tendency.

Creeping effects in distillation, already mentioned on page **48**, are sometimes given extra consideration in the construction of columns and still heads. The exit tube is sharply bent, perhaps, or the exit tube is inserted through the column so as to project inside the apparatus. The intention in each case is to provide a convexity of inside surface, if not already provided as in the pear shaped columns, which will help in the eradication of creeping

Preparation of Ethyl Benzene

effects. *The reflux ratio* is the ratio of the amount refluxed back to the amount allowed to distill over. Columns and flasks are often provided with a cold point, near the exit tube, which serves the purpose of refluxing back into the column any desired proportion of the material passing through the column. The cooling point is so arranged that the fraction condensed back can be estimated by comparing the number of drops falling from cooling point and exit tube respectively. A simple cooling point is illustrated above the Hempel column and the term reflux ratio usually refers to some such cooling point in the distilling apparatus near the exit tube. The ratio can be varied by changing the rate of water flow in the cooling point and is useful in standardizing the description of distillation processes.

Part II

Additional Exercises

Exercise 8

Preparation of Veronal

(17, 23)

VERONAL is a hypnotic which has been extensively used in medicine. It is the ureide of diethyl malonic acid and is usually prepared by the condensation of the diethyl ester of diethyl malonic acid with urea in the presence of sodium ethoxide.

$$(C_2H_5)_2C(COOC_2H_5)_2 + NH_2CONH_2 \rightarrow$$
$$(C_2H_5)_2C-CO-NH + 2C_2H_5OH$$
$$CO-NH-CO$$

The diethyl ester of diethyl malonic acid (often simply called diethyl malonic ester) can be prepared from malonic ester (diethyl malonate) by the use of ethyl bromide and sodium ethoxide. The equation can be written as follows:

$$2C_2H_5Br + Na_2C(COOC_2H_5)_2 \rightarrow (C_2H_5)_2C(COOC_2H_5) + 2NaBr$$

The Literature

This can be summarized briefly. From malonic ester the diethyl malonic ester is prepared by adding the theoretical quantity or a slight excess of ethyl bromide to a mixture of malonic ester and two molecular proportions of sodium ethoxide in a convenient quantity of absolute alcohol, with stirring, and subsequent refluxing until the reaction is at an end. The end point is marked by the conversion of all the sodium ethoxide, or sodio-malonic ester, into sodium bromide. Since sodium ethoxide and sodio-malonic ester are hydrolyzed on contact with water to give sodium hydroxide, a test portion of the solution, when diluted with water, shows an alkaline reaction to litmus paper unless the reaction is at an end. The usual time needed for refluxing is about four hours. When neutral, the alcohol is distilled off, water added to the residual mixture, and the crude diethyl malonic ester extracted with ether. The ether solution is dried and the ether removed by distillation. The residual crude diethyl malonic ester is distilled. A small proportion of unchanged malonic ester comes over first and then some mono-ethyl malonic ester. A crude diethyl malonic ester collected from 212°C.-224°C. is sufficiently pure for the preparation of veronal.

Veronal is prepared from diethyl malonic ester by admixture with the theoretical quantity of urea and an alcoholic solution of sodium ethoxide containing about 10 per cent of sodium ethoxide. The alcohol is present in roughly three times the amount of the ester. The mixture is either heated in an autoclave at 100°C. or near that point for four or five hours or refluxed vigorously for seven or eight hours or more. The end point cannot be gauged by test. Most of the alcohol is removed by distillation, the residue cooled and diluted with cold water and immediately made faintly acid with hydrochloric acid. Veronal is quite soluble in alcohol, less than 1 per cent in water, and readily hydrolyzed by alkalies. It melts at 191°C. (corr.). It can be recrystallized from dilute alcohol (50 per cent to 70 per cent in water). The yield depends on the temperature and time involved in the condensation and is markedly affected by the presence of water in the alcohol. The operation should therefore be performed with a very dry alcohol and moisture should be excluded. For the same reason the addition of water at the end of the condensation should be followed immediately by an acid to prevent hydrolysis. The melting point quoted is a corrected melting point.

Procedure

Using 50 g. of malonic ester the required amount of sodium would be almost 15 g. and the weight of ethyl bromide required would be 70 g. Since ethyl bromide is very volatile (b.p. 39°C.) it would be well to use 75 g. of the bromide to allow for losses in handling and in incomplete condensation when refluxing. The convenient quantity of alcohol to dissolve 15 g. of sodium without deposition of sodium ethoxide when cool would be something over 100 g., but 200 g. would be recommended to allow for solution of the malonic ester. Even with this excess, however, it will be seen that the addition of malonic ester to the solution of sodium ethoxide in alcohol will be followed by the formation of a soft gelatinous mass of sodio-malonic ester, and in order to make certain of the uniformity of the mass it would be necessary to heat on the water bath until liquid and then allow to cool. The ethyl bromide is then to be added in portions into the stirred cold mass which slowly reacts giving sodium bromide and a more mobile mixture of reaction products. When the mass is so mobile that the ethyl bromide can be assumed to be in thorough contact with the soft solid remaining (this usually means stirring for an hour or more), the mass can be carefully refluxed on the steam or water bath. It would be unwise to add without stirring simply because the ethyl bromide would react so slowly that when brought to reflux later the reaction might proceed to a point of good admixture with a large excess of ethyl bromide still present. If then the reaction

Additional Exercises

proceeded quickly the reflux condenser would perhaps be unable to condense the ethyl bromide being evaporated at a fast rate by the heat of the reaction and the steam bath together.

Control of the Heat of Reaction

When an exothermic reaction, which is the usual reaction in synthetic organic chemistry, is to be controlled, it is necessary that no part of the procedure should involve the operator in a situation where the means of control are inadequate when slight variations of technique occur. The speed of reaction always increases with increasing temperature—this means, in exothermic reactions, a tendency to explosive violence in all cases though most cases are easily controlled. For this reason, instructions often call for the addition of one of the reagents in small quantities at a temperature sufficient to promote reaction at a rapid rate but controllable by the cooling devices when small quantities are used. After a certain interval of time it is then safe to add another small quantity of reagent. If the reaction mass is not homogeneous it is important to note descriptions of the means of stirring, since the rate and the control of the reaction then also depend on another factor of surface contact which can be enormously varied by alteration of the efficiency of stirring (see Part III).

We must arrange to prepare the solution of sodium ethoxide, to add the malonic ester, then the ethyl bromide while stirring, and finally to reflux. Since water is undesirable and some of the materials are hygroscopic, it would be desirable, though not absolutely necessary, to perform the whole operation without transfer and under protection from the moisture of the air. It would be proper to arrange a flask surmounted by a condenser, to introduce the absolute alcohol into the flask, and to introduce the sodium into the alcohol piece by piece through the condenser or by detachment of the flask. If a bulbed condenser is chosen, for good cooling later when using ethyl bromide, the sodium cannot be dropped through the condenser for fear of blocking. The heat of the reaction will probably cause some alcohol to boil but this is refluxed back by the condenser.

Absolute Alcohol

For this part of the procedure the "absolute" alcohol is the more usual absolute alcohol of the laboratory containing $\frac{1}{2}$ per cent to 1 per cent of water. This can be prepared from 95 per cent alcohol by mixing with freshly broken lumps of lime from $\frac{1}{2}''$ to $1''$ in diameter in such quantity that it comes to the surface of the alcohol in a round-bottomed flask. Under a reflux condenser surmounted by a calcium chloride tube the mixture is brought to the boil on a steam bath. If the quantity of alcohol is large (two litres or more), the refluxing may be violent as the lime begins to crumble; for larger quantities it would be best to allow to stand overnight before refluxing. Reflux for an hour after the lime has mostly disintegrated to a powder and distill into a side-arm flask with a calcium chloride tube attached.

A more "absolute" alcohol, containing less than one part of water in a thousand, can be obtained from this dried alcohol by dissolving in a litre of the alcohol 7 g. of sodium and adding subsequently 25 g. of ethyl phthalate and refluxing for one hour. The water is removed by the reaction $RCOOC_2H_5 + C_2H_5ONa + H_2O \rightarrow RCOONa + 2C_2H_5OH$. Subsequent distillation in apparatus carefully protected from moisture gives a very dry alcohol which is very hygroscopic. Ethyl phthalate present in excess boils at 222°C. and is thus not distilled over with the alcohol.

For stirring the mass after the addition of malonic ester, a mechanical stirrer could be fitted to run through the condenser or the flask could originally be fitted with an addition tube with a condenser attached to

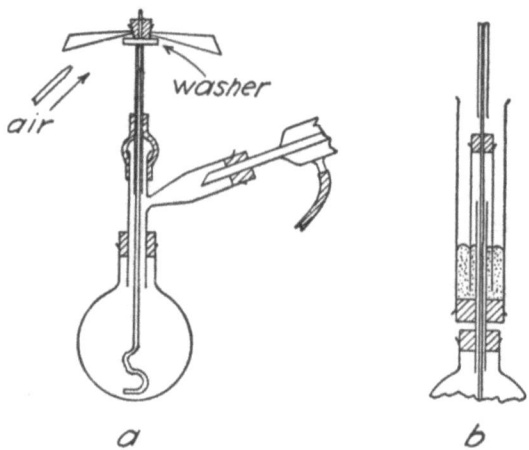

Fig. 15. (a) An air blast stirrer made from a cork and the top of a tin can cut into vanes. The supporting tube is attached to the addition tube by rubber tubing and supported in a clamp. (b) A mercury seal for a stirrer leading into a closed flask. The stirrer is supported here as in (a).

the side arm. Figure 15a shows a simple stirring device which could be used with a blast of air. This would be suitable for this exercise though not providing absolute protection from atmospheric moisture. Figure 15b shows a mercury seal which can be made easily and fitted for a vane blast stirrer or a motor stirrer.

The ethyl bromide should be added from a dropping funnel in the course of an hour and, the mass being now all mixed and mobile, the stirring discontinued. The mixture is then refluxed gently for four

Additional Exercises

hours by which time a test drop in water should be neutral to litmus. If the mixture is much delayed in reaching the neutral point, some more ethyl bromide should be added. The flask is then connected with a bent tube leading to a condenser and most of the alcohol distilled off. Water is added to the residue and the mass extracted with ether. The ether extract is dried with a suitable drying agent, the ether removed, and the residue distilled from a low side-arm flask with an air condenser. The portion boiling from 212°C.-225°C. (on the usual uncorrected thermometer from 208°C.-220°C.) is collected and weighed. This is used in the preparation of veronal as crude diethyl malonic ester, the quantities appearing below being amended according to this yield.

In the condensation of urea with diethyl malonic ester the best available "absolute" alcohol is used. The sodium ethoxide is prepared as before and the mixture of diethyl malonic ester and urea in another quantity of alcohol added. The correction of a thermometer is described below.

Correction of a Thermometer for Melting Points

Unless there is available a thermometer which has been standardized by some reputable agency such as the U. S. Bureau of Standards and is made of a normal glass (a glass whose behavior on heating and cooling is reproducible over a long period of time), and unless such a thermometer is used with the whole mercury column immersed in the bath liquid, a certain correction must be made in the observation of the melting point.

Usually the most serious error is that introduced by the non-immersion of the whole mercury column in the bath liquid. This error can be eliminated for the most part by adding to the observed reading a correction of .000154 $(t - t^1)$ N degrees where t is the observed temperature, t^1 is the average temperature of the portion of mercury column not immersed (it can be gauged roughly by a thermometer hanging alongside the melting point thermometer at a point near the middle of the portion of mercury column not immersed), and where N is the number of degree intervals in the portion of mercury column not immersed. The figure .000154 is the apparent coefficient of expansion of mercury in glass. Certain short-scale thermometers are available in the market which can be used with the whole column of mercury immersed in the bath and this correction can thus be avoided.

Unless the thermometer provided is of guaranteed behavior, whether long scale or short scale, it is necessary for the operator to check the accuracy of the thermometer before submission of results for publication. Such a check on the thermometer would be provided by examining the behavior of known pure substances in the melting-point apparatus under the usual conditions. Such an examination would also dispense with the necessity for a stem correction if carried out in the following way. A pure specimen of a material of known true melting point is introduced into the apparatus and its melting point observed. The difference between the true melting point and the observed melting point is noted. Other specimens of materials of known true melting point are also introduced, to the total of seven or eight, so as to cover the range from 0°C. to 360°C. in approximately regular intervals; in each case the difference between the observed and true melting point is noted. If now the results are plotted so that one co-ordinate shows corrections to be applied and the other shows observed temperatures, then the curve obtained by joining the plotted points gives an estimate of corrections to be applied in intermediate temperatures for use when dealing with new substances or with substances not melting at the observed points. Such a correction chart for the thermometer, while not highly accurate, is good enough to justify the use of the notation X°C.(corr.) when reporting in the usual way. It is still subject to all the error of the apparatus. Substances suitable for use as standard pure substances are found in the list below. They should be obtained in the highest possible purity from market sources and then crystallized from suitable solvents two or three times, or sublimed to constant melting point.

>P-toluidine 45°C.
>Diphenylamine 54°C.
>Naphthalene 81°C.
>Benzoic acid 122.5°C.
>Salicylic acid 160°C.
>Anisic acid 184°C.
>Anthracene 216°C.
>Anthraquinone 285°C.

Correction of a Thermometer for Boiling Point

The correction of a long-scale thermometer by the use of the formula given above for the portion of mercury column not immersed, or the use of pure substances of known boiling point for the construction of a correction chart, is the common procedure. Substances suitable for use in making the chart of corrections can be found in the following list. The use of the chart should be limited to the size of distilling flask used in the estimation of the correction and the rate of distillation (usually one drop a second) at which the correction was observed. Since superheating is very apt to occur with small quantities of liquid, the notation of boiling point should be made while there is still a residue of five to ten cc. of liquid in the flask and the distillation should be kept up until the thermometer is steady. The use of a small piece of porous tile and a rotating small flame would be advised.

>Chloroform 61°C.$_{760}$
>Benzene 80°C.$_{760}$
>Water 100°C.$_{760}$
>Ethylene dibromide 131°C.$_{760}$
>Aniline 184.5°C.$_{760}$
>Nitrobenzene 211°C.$_{760}$
>Quinoline 237.5°C.$_{760}$
>Benzophenone 306°C.$_{760}$

A correction for the barometric pressure at observation, negligible in the case of melting points, must be made for boiling points if the pressure at ob-

Additional Exercises

servation is not 760 mms. The correction to 760 mms. is .00012 (760 − p) (273 + t) for most liquids and .00010 (760 − p) (273 + t) for water and alcohols where p and t are the observed barometric pressure (corrected as described on the barometer for the temperature of the atmosphere) and temperature of distillation in degrees centigrade respectively (Young, *Fractional Distillation,* Macmillan and Co., 1903) (distillation, pp. **82, 84, 107**).

Exercise 9

Preparation of Ethyl Acetate

(6, 23, 35, 79)

The Literature

The formation of an ester from an acid and an alcohol is a common operation and therefore finds a place in the textbooks of laboratory methods (see Part IV). In Houben Weyl, *Methoden der Organischen Chemie,* there is a description of most of the methods used in the esterification of acids and a particular reference to ethyl acetate in these terms: "a mixture of equal volumes of alcohol and concentrated sulphuric acid is heated in an oil bath to 140°C. At this temperature a mixture of equal volumes of alcohol and acetic acid is allowed to run in at the same rate as the ester distils over." Further reference to laboratory manuals and the Journals establishes the procedure as a simple example of continuous dehydration by the sulphuric acid to give a distillate composed of the ester, acetic acid, water, alcohol, and traces of ether. The reaction is, in the main, $CH_3COOH + C_2H_5OH \rightarrow CH_3COOC_2H_5 + H_2O$ and the distillate contains traces of ether from the side reaction $C_2H_5OH + C_2H_5OH \rightarrow C_2H_5OC_2H_5 + H_2O$ which we would expect under the circumstances. The amount of the mixture of acetic acid and alcohol to be added to the mixture of sulphuric acid and alcohol determines the yield and can be any convenient quantity. The amount of sulphuric acid can be any convenient quantity and is effective for many times its amount of added mixture of acetic acid and alcohol. Theoretically it should be effective for an unlimited quantity of acetic acid and alcohol since it does not distill below 300°C. and acts as a dehydrating catalyst at a temperature which allows for continuous distillation of both reaction products. The alcohol to be used need not be absolute.

Procedure

The oil bath (cottonseed, rapeseed, olive oil, hydrogenated kitchen oils such as "Crisco" but not paraffin wax or kerosene oils since these are more inflammable) is easily arranged with a distilling flask or an ordinary flask having a bent-tube connection with a condenser. A thermometer should hang in the oil bath and the oil, being viscous, should be frequently stirred with a glass rod or the thermometer in order to ensure

Additional Exercises 101

a more or less uniform temperature in the bath. A dropping funnel, with its end below the surface of the reaction mixture, would permit addition of the alcohol-acetic acid mixture in a measurable way into the body of the reaction mixture. Remembering that sulphuric acid generates heat when mixed with ethyl alcohol the admixture should be made carefully. The highest boiling component in the distillate being acetic acid of b.p. 118°C. a water condenser is permissible.

The properties of the reagents and products here mentioned for the first time are the following. Ethyl acetate b.p. 77°C., slightly soluble in water (about 6 per cent), sp.gr. 0.9, hydrolyzed readily by alkalies, but not appreciably when washed with cold solutions of sodium carbonate and afterwards with water, soluble in alcohol and most organic solvents. Acetic acid b.p. 118°C., very soluble in water and also in most organic solvents, sp.gr. 1.05.

The distillate, containing not only ethyl acetate and water but also acetic acid and alcohol which are both soluble in ethyl acetate and water and in each other, may not separate into two layers. In any event it would be proper to add concentrated sodium carbonate solution to neutralize the acetic acid as evidenced by cessation of evolution of carbon dioxide or, better, by test with a moistened litmus paper. The sodium carbonate is used as concentrated solution, or even as a solid, in order to keep the quantity of water at a minimum, since ethyl acetate is slightly soluble therein. Two layers should now appear because one of the substances making for miscibility (acetic acid) has been removed and because the presence of sodium acetate and sodium carbonate in the water should diminish the solubility of the ethyl alcohol, ethyl acetate, and ether in the water.

This effect is called a "salting out" effect wherein the solubility of a substance in water is diminished by the solution of salts in the water. This effect is often turned to advantage in the separation of organic substance somewhat soluble in water by the addition of an inorganic salt (usually sodium chloride) to the water solution or to the mixture of the wanted material and water. Thus in the preparation of aniline (Exercise 14), common salt is added to the water solution before extraction with ether and commonly the sodium salts of sulphonic acids are precipitated from water solution by the addition of sodium chloride or strong brine.

Organic substances are in general insoluble in water but substances containing oxygen in the form of hydroxy, carbonyl, carboxyl or ether groups are likely to be soluble in water if the percentage of oxygen in the molecule is something over 25 per cent. This is also true of nitrogen in the form of amine, amide, and oxime groups. Organic substances are also in general soluble in each other but those organic substances which show great water solubility are unlikely to be good solvents for organic substances insoluble in water: like dissolves like. An extreme case is represented by the salts of organic acids which resemble closely the typical inorganic salt in being soluble in water and quite insoluble in hydrocarbons and the majority of other organic substances.

The typical intermediate in solubility is such a substance as ethyl alcohol having about half its molecular weight as oxygen in a hydroxyl group. It is soluble in water and yet is a good solvent for most organic compounds (solubility, pp. **33, 40, 155**).

The water layer should now be removed and the ethyl acetate layer washed once more with a little water or salt solution. The alcohol remaining in the ethyl acetate must be completely removed since it boils at 80°C. (b.p. ethyl acetate 77°C.) and cannot be separated from ethyl acetate by distillation. Since calcium chloride combines with alcohol and water, it should be used, as a solid or even as a very concentrated solution, to remove the alcohol and water. The ester layer is then separated and distilled. The quantity of ether in the acetate is negligible after distillation.

It sometimes happens that a product forms a constant boiling mixture with a solvent such as ether used in its preparation. When the solvent cannot be washed away with water (ether, benzene, and chloroform, etc., etc.), it is necessary to fall back on other methods of separation. Mention has already been made of the fact that cold concentrated sulphuric acid can sometimes be used for removal of solvents containing oxygen in the molecule. When this cannot be employed there are still other substances which may not be mentioned in laboratory manuals but which can be found in the literature to cover the case in hand. For ether separations, for instance, mention has been made of phosphoric acid, aluminum bromide, ferric chloride as possible agents for removal.

Exercise 10

Preparation of Acetamide
(8, 17, 25)

From Ammonium Acetate

In a Ladenburg distilling flask, heat a mixture of ammonium aceta and glacial acetic acid, or ammonium carbonate and acetic acid, in abo a triple or quadruple proportion of the acid. Lead the distillate awa with a short air condenser or into a distilling flask attached directly the side arm as a receiver. Arrange the heating so that a very slow di tillation occurs, the temperature of the thermometer being at 102°C 104°C., and continue for 3½ hours. Then slowly let the temperature the distillation reach 150°C. Reject the distillate. The distillation is be continued so as to collect the fraction from 150°C.-210°C. and a other, mostly amide, around 220°C. Use the proper condensers and, the amide solidifies and clogs the apparatus, warm to melt it. Redist the fraction 150°C.-210°C. and collect therefrom the portion distillir above 215°C. and add it to the crude amide. Crystallize from benzer or chloroform or by the slow addition of ether to a concentrated solutic of the amide in methyl alcohol. Test out the crystallization procedur before proceeding with the bulk crystallization, noting concentration b.p. 222°C. m.p. 82°C.

From Ethyl Acetate

If ethyl acetate is mixed with two parts of a concentrated aqueou ammonia and allowed to stand overnight, the mixture becomes homogene ous and, on distillation in the same fashion as before, acetamide can b obtained. Do not lose ammonia from the mixture while standing, an arrange to dispose of the ammonia on distillation. Decide on the neates

Exercise 11

Benzoic Acid from Ethyl Benzene
(17, 36, 56)

The Literature

This reaction has not been extensively discussed and the instructions are such as would be found in the literature in a form suitable for the more advanced student. A précis follows:

"Ethyl benzene is converted in fair yield into benzoic acid by long heating with dilute nitric acid (1 vol. of nitric acid of sp.gr. 1.4 with 2 vols. of water). The acid is filtered, freed of nitrobenzoic acids by treatment with tin and hydrochloric acid, and then dissolved in sodium carbonate. It is thrown out once more with hydrochloric acid and obtained pure in one crystallization from boiling water. Melts at 119°C.-120°C. and gives no depression of melting point when mixed with a specimen of benzoic acid melting at 119°C.-120°C."

Procedure

Reviewing the properties of ethyl benzene, we find it to be insoluble in water, b.p. 136°C., soluble in alcohol and ether. Benzoic acid melts at 122°C., is slightly soluble in cold water and more soluble in hot (about 5 per cent), and is soluble in alcohol and ether.

It is first necessary to examine the instructions for an understanding of method. It seems that the liquid ethyl benzene is gradually transformed into solid acid which will make its presence evident by precipitation when cold even if dissolved when hot. We can presumably use any desired quantity of dilute acid and the reaction must obviously be carried out under reflux. Assume nitric acid to have the minimum capacity for oxidation in the decomposition $2HNO_3 \rightarrow H_2O + 2NO_2 + O$ for the oxygen required in the decomposition $C_6H_5CH_2CH_3 \rightarrow C_6H_5COOH + CO_2 + 2H_2O$, and work with something over 10 moles of nitric acid in a trial run.

No end point is mentioned, but since ethyl benzene is a liquid, an end point is suggested when the cool mass shows no oily drops of insoluble ethyl benzene on the surface or, alternatively, when the deposition of solids on cooling seems to be at a maximum.

The crude acid is filtered. Treatment with tin and hydrochloric acid

Additional Exercises

obviously aims at the conversion of nitrobenzoic acids to aminobenzoic acids by reduction. These aminobenzoic acids are amphoteric and the instructions assume that the reader knows how to separate aminobenzoic acids (ortho, meta, or para) from benzoic acid completely. Being amphoteric, and forming salts with acids as well as bases, the instructions probably mean that the crude acids are to be treated with tin and hydrochloric acid and filtered from the aminobenzoic acids remaining in solution in the hydrochloric acid. Reference to the literature confirms the solubility of the amino acids in hydrochloric acid and the instructions are now translated into the following: the crude acids are mixed with an equal volume or more of strong hydrochloric acid and tin is added in excess and the mixture brought to reflux for some time (this being an approximation to a reduction of nitrobenzene or a nitrobenzoic acid from laboratory manuals). We can here describe no definite end point and the period of reflux is again decided as an hour or so by comparison with similar procedures. On completion of the reflux, the mass should be cooled and filtered and the residue of benzoic acid and tin washed with cold water.

The amino-acids being now removed, the benzoic acid can be separated from the tin by treatment with a solution of sodium carbonate or sodium hydroxide which dissolves the acid as the sodium salt. The water solution, if clear and colorless, is treated with hydrochloric acid to precipitate benzoic acid. This is crystallized in the usual way.

Make careful notes for comparing yields with other operators and observe a mixed melting point with a specimen of benzoic acid from other sources. The mixed melting point of identical substances should show no depression, but since impurities in general lower the melting point of a substance it would be expected that the product, if it were not benzoic acid, would behave as an impurity in the benzoic acid and so lower the melting point. This mixed melting-point test, together with analyses and other observations, is much used in proof of structure.

The validity of this test of the identity of a substance obviously depends on the accuracy of the observation and its relation to the expected lowering by admixture with impurity. Other factors also need consideration at this point.

In the first place, the impurity must be to some extent soluble in the liquid pure material. Sand would not lower the melting point of most organic substances. That this is the case can be seen by considering the equilibrium at the melting point of a pure substance. The change of melting point by addition of a foreign substance is related to the change of the vapor pressure of one of the components by this addition. This change would not take place if the added substance were absolutely insoluble in the pure material since the vapor pressure of mutually insoluble substances is merely the sum of the vapor pressure of both at that temperature. It is true that very few liquids are absolutely insoluble in each other; but sand, for instance, is quite insoluble in most organic substances.

It is also necessary that thorough admixture of the specimens be made by grinding in a mortar or by crystallization together or by melting together, allowing to solidify, and re-melting.

The test is sometimes done repeatedly with varied proportions of the two materials in order to reduce the chances of error through compound formation between two individuals. If such compound formation took place in a certain stoichiometric proportion which was close to the proportion accidentally chosen, and this compound melted at the same point as both individuals, a wrong conclusion would be drawn from the test unless repeated with a different proportion of materials.

When the impurity is soluble it affects the vapor pressure of the liquid component more or less in accord with the law that the partial vapor pressure of a component varies as its mole fraction, so that the greater disturbance is effected by the greater amount of impurity. The amount of that disturbance, as we recall, can even be used to estimate the molecular weight by the method of the lowering of the freezing point. The amount of lowering caused by the solution of one gram-molecule of a substance in 100 g. of solvent (the molecular depression) is, however, a number peculiar to the solvent; cases are occasionally reported where admixture with a large amount of impurity has escaped detection in the usual apparatus.

The most significant point for observation in taking melting points is the point of final clear melt though the range of melting point, when more than ½°C. to 1°C., is also an indication of the presence of an impurity. If the ordinary apparatus is used for an estimation of the lowering of melting point by the addition of impurity, it is this point of final clear melt which is used for comparative purposes.

Some substances show such a large molecular depression that this apparatus can actually be used for the estimation of molecular weights in the place of the usual accurate Beckmann apparatus. Such a method is described by Rast (*Ber.* **55**:1051, 3727) and is illustrated by J. H. C. Smith and W. G. Young (*J. Biol. Chem.* **75**:289). About 0.2 g. of the substance is weighed accurately in a test tube and covered with approximately ten times its weight of camphor and the mixture again accurately weighed. The mixture is heated in a test tube with a yellow flame until completely melted and is then allowed to cool with stirring until solidified (some operators prefer to remove sublimed camphor). A little of the mixture is placed in a melting-point tube usually of a little larger diameter than the average capillary tube, and packed with a very fine rod of glass. The melting point is taken, as stated before, to be the point of final clear melt and is compared with the same point for pure camphor. The molecular weight of the substance then is calculated from the statement that the molecular lowering (see above) for camphor is 397°C. (melting point, pp. **36, 39, 151**).

Exercise 12

Preparation of Brombenzene

(17, 26, 35)

ALLOW 20 cc. of benzene to react with 40 grammes of bromine with an iron nail or a gramme or two of iron filings as a catalyst. Warm if necessary to start the reaction. Dispose of hydrogen bromide. Let stand overnight. Add water and decant twice, then wash in a separatory funnel with appropriate material for removing acids, finally with water to remove traces of this material. Dry with a drying agent. Distill. Collect between 140°C.-170°C. and fractionate, collecting as product the material b.p. 154°C.-156°C. The residue can be crystallized from alcohol, decolorizing if necessary with a charcoal, and is p-dibrombenzene m.p. 89°C., b.p. 219°C.

The fractional distillation of completely miscible substances usually approximates the distillation of an ideal mixture when the components are of the same chemical type. The mutually insoluble liquids give, at a certain temperature, their own vapor pressures at that temperature when mixed so that the boiling point is lower than that of either component. However, the ideal mutually soluble liquids (the ideal mixture so-called) contribute at a fixed temperature not their full vapor pressures independently but this value multiplied by a fraction representing the mole fraction of each component in the mixture (Raoult's Law, Henry's Law). It is an easy matter to calculate the boiling point of ideal mixtures from a record of the vapor pressures of the pure components and a knowledge of the mole percentages of each component in the mixture.

The equation representing the contribution of each component would be of the form $F(A) + (1 - F)B = V.P.$ where F is less than 1 and A and B are the vapor pressures of the components at that temperature. It follows, putting the equation at boiling in the form $F(A - B) + B = 760$, that A must be above 760 and that B must be below 760 and that therefore the boiling points must lie between the boiling points of the two constituents. Deviations from the law, however, have already been encountered in the case of azeotropic mixtures. The deviation is called positive if the vapor pressure is higher than anticipated from the law, and negative if lower. Azeotropic mixtures of positive deviation are often encountered in organic chemistry but rarely those of negative deviation which give maximum boiling points (distillation, pp. **82, 84, 153**).

Exercise 13

Preparation of Nitrobenzene

(25, 35, 79)

CAREFULLY mix equal chosen volumes of concentrated nitric acid and concentrated sulphuric acid, cooling under the tap. From a dropping funnel, add the mixture carefully (in the course of an hour for quantities below 100 g.) to three-quarters of this chosen volume of benzene, using a hood to remove the fumes or, if not available, some inverted funnel connected to a pump or draft pipe or other arrangement not involving the use of rubber or cork in the fumes. The vessel must be frequently shaken by swirling, and frequently cooled under the tap, keeping the temperature below 50°C. When all is added, warm on the steam bath to keep at 50°C. for a half-hour or more, swirling frequently. An end point is reached when a drop of material sinks in water, the nitro compounds being denser than water, and benzene less dense. Dilute with much water, wash away acids (the nitro compounds are stable to alkalies) and finally with water, separate, dry, and decant from the drying agent into a small distilling flask with a low side arm, the boiling point of nitrobenzene being high (211°C.). The drying agent can be washed if desired with a low-boiling solvent and the washings added. Cotton wool or glass wool can often be used, plugged gently, as a filtering device. Distill with a yellow moving flame and do not collect higher than 212°C. on the usual uncorrected thermometer, the residue being liable to violent decomposition at high temperatures. b. p. 211°C. (corr.)

This description applies to specimen quantities. If the volumes are greatly increased it is important that the addition of the nitric acid should be made in the course of a much longer time. Nitrations are strongly exothermic reactions and are usually performed on substances not miscible with the nitrating agents. In performing such operations the addition is made slowly and the temperature kept near the maximum temperature quoted (in this 50°C.), with good mixing whether so advised or not. This ensures a reaction speed at least as fast as the quoted one. The multiplication of quantities is not to be done recklessly: for the nitration of 1,000 g. of benzene, for instance, the procedure would first be repeated, using double the time or more, on 200 g. Then 400 g. would be tried at the same rate of addition of nitrating agent in the course of five or six times the number of hours, with careful note of any tendency of the reaction mixture to run above the maximum temperature, etc., etc. Safety depends on having at no time sufficient excess of unreacted nitrating agent to initiate an uncontrollable rate of reaction.

Exercise 14

Preparation of Aniline from Nitrobenzene
(23, 25, 27, 35, 37)

The Literature

The reaction has been described in many laboratory manuals in such terms as the following:

Place in a large flask a mixture of 90 g. of mossy tin and 50 g. of nitrobenzene. Add 200 cc. of conc. HCl in portions of 20 cc. with good shaking and keep the reaction near the boiling point (an air condenser or water condenser should surmount the flask for this reason). Cool in water if the reaction threatens to become too vigorous. After the acid has all been added keep on the steam bath for an hour or so until no oily drops of nitrobenzene can be seen. If necessary the exercise can be interrupted at this point.

Attach the reaction flask for steam distillation and steam distill after making the mixture strongly alkaline with conc. sodium or potassium hydroxide. Continue the steam distillation past the point of non-appearance of oily drops or cloudiness in a test portion of the distillate until a few hundred cc. more of distillate have been collected. Add sodium chloride almost to saturation in the water (about 30%) and extract twice with about half the volume of ether. Dry the ether solution overnight with solid potassium hydroxide or sodium hydroxide. Decant. Drive off the ether. Distill the residue and collect at 180°C.-184°C. Colorless oil, sp. gr. 1.024 at 16°C., slightly soluble in water, darkening in the air without much decomposition.

Procedure

Some texts recommend that at the end of the reduction the reaction mixture should be cooled and extracted with ether to remove unchanged nitrobenzene and any other compounds which do not combine with the hydrochloric acid. The aniline remains in the water solution as aniline hydrochloride.

Other reagents commonly employed for the reduction of nitro-compounds are stannous chloride in water or alcohol-water mixtures together with hydrochloric acid, or iron with dilute acetic acid. When dealing with amines which are not very volatile in steam the isolation of the free amine from the complex salt formed with stannic chloride and hydrochloric acid is apt to be troublesome. Extraction with ether after decomposition of the salt with very concentrated sodium hydroxide sometimes succeeds though emulsions are apt to form. An alternative is to treat with hydrogen sulphide and filter off the tin sulphide but this is a process that sometimes requires good washing of the sulphide.

Tin and HCl as reducing agents are liable to give chlorinated amines as by-products particularly in drastic conditions of long heating and the use of concentrated acid: otherwise the operation is very general and dependable.

Exercise 15

Preparation of P-tolunitrile

(13 or 14, 27, 29, 35, 79)

The Literature

The common method of preparation is by means of the Sandmeyer reaction on p-toluidine. It is described in many laboratory manuals and in the *Journal of the American Chemical Society* (**46**:1001).

A solution of p-toluene diazonium chloride is prepared and added to a hot solution of cuprous cyanide, freshly prepared, in such manner as to provide good mixing before decomposition of the diazonium chloride solution. Since the diazonium compounds generally are decomposed by hot water to give phenols and by-products such as azo compounds, biphenyls, etc., the usual conditions call for concentrated solutions, temperatures which have been observed to give the minimum of by-products, and good admixture. In individual cases, or in some classes of compounds, certain details are added which have been found important for that individual or class but not for all classes. Thus it is sometimes necessary to avoid even slight excesses of nitrous acid, the excess being removed by a current of carbon dioxide (or the use of dry ice) or nitrogen or other inert gas, whereas in general a slight excess is neglected. This is not unexpected and the information only confirms us in the rule of initiating no new procedure or synthesis without a thorough examination of the literature even for the most common reaction. b.p. $217°C._{760}$ (corr.) $91°C.$; m. p. $29°C.$

The nitrile formed is removed by steam distillation and is washed with dilute cold sodium hydroxide solution to remove p-cresol if it is not solid in the receiver when cold. It may even need redistillation in vacuo.

Note the discussions of stirring and the use of a benzene layer. Note the potassium-iodide starch paper test for nitrous acid. Note the amount of ice considered to be sufficient for keeping the reaction cool in the ordinary way of diazotization. Diazotizations are usually done near $0°C.$ but sometimes at room temperatures. The amine nitrate is often used when the other salts are sparingly soluble. Sometimes the amine salt is dissolved hot and allowed to crystallize with stirring and the diazotization carried out on the suspension of fine crystals. On occasion the diazo solution is filtered, or it is made alkaline or neutral for the introduction of certain groups such as the arsonic acid group or for coupling with phenols.

Procedure

The cuprous cyanide is prepared in a hood because cyanogen is evolved in the reaction of copper sulphate with potassium cyanide. The solution should be clear and not dark.

The decomposition of the diazo compound by the cuprous cyanide solution should also be done in the hood.

Exercise 16
Preparation of Hydrocinnamic Acid
(3, 17)

THIS INVOLVES the preparation and use of a 2.5 per cent sodium amalgam (*J. Am. Chem. Soc.*, **45**:1740; Gattermann, *Practical Organic Chemistry*, [1933], 225). The end of the reaction is tested for by the addition of dilute potassium permanganate, and crystallization from water. (See also Fisher, *Laboratory Manual of Organic Chemistry*, Wiley & Sons, 1920.)

Exercise 17
Preparation of Acetanthranilic Acid from Acet-o-toluidide
(14, 41, 42)

THIS PREPARATION involves: an oxidation with neutral permanganate; a test for the end point; the destruction of the excess permanganate; hot filtration. Details should be obtained from the literature rather than from laboratory manuals.

Exercise 18
Preparation of Quinoline
(13, 25, 26, 27, 35, 37)

THE PREPARATION of quinoline by the Skraup method is often uncomfortably violent. Modifications in the synthesis have been described. An acetic acid modification, and a boric acid modification, initiated on the presumption that the acetate or borate of glycerol would only slowly liberate the glycerol for reaction, are apparently very satisfactory. A discussion is given by Cohn, et al, in *J. Am. Chem. Soc.*, **52**:3685 and **50**:2709. The preparation involves: refluxing for a day or two; steam distillation; further steam distillation; removal of aniline and phenol by diazotization followed by steam distillation, extraction, distillation.

Part III

Individual Work

Individual Work

IN THE succeeding pages some more advanced experiments are described. In no case is the description detailed and in some cases there is merely an indication of source material in the literature. When a maximum yield is desired, or a convenient process, laboratory texts should be consulted or such convenient collections of syntheses as *Organic Syntheses* referred to in the chapter on the literature of organic chemistry. We will assume that some experience has already been gained in the handling of solvents and that some good habits have been formed.

It should now be obvious that no filtrate or residue is thrown away until the material sought has been obtained in the yield desired, and its identity established: this because the solubility of the material may be quite different from that claimed in the directions, owing to a difference in yield or nature of by-product. It should have been realized that a dependence on literal instructions is useless and dangerous: the less thinking required, the less easy becomes the habit of thinking. It should be clear that no assumption of the approximate amount of impurity can be made from a boiling range: a range of three degrees may mean one per cent of impurity or 20 per cent of impurity, depending on the nature of the impurity. Thorough washing should be almost automatic in every filtration, and the operator is expected to have this in mind.

The question of yield is sometimes important. One may be required to try, in empiric fashion, to increase a yield. This resolves itself into a process of varying one factor of temperature, solvent, or the like and observing the result; then comes a variation of another single factor with a like observation; then comes an examination of simultaneous variation of the two factors. If any general theory of reaction is helpful in choosing the most important factor, so much the better. If it is easier to test for increase or decrease in a major by-product than for yield of product, it may be found possible to use that test as a counter-indication of yield. If it is possible to analyse the whole reaction product by distillation or crystallization, the analysis may well be undertaken since an observation of the nature of the by-products may lead to rational procedure for their elimination. For the rest, there is only a matter of examining filtrates, residues, for any unrecovered material.

It is necessary, also, to learn a respect for the historical value of the literature. There is much illustrative comment hidden away in obscure places. When at fault, we may be in the same difficulty as a writer of long ago working on a problem analogous to ours or with material somewhat like ours. We must, however, retain a wariness concerning the accuracy of the literature. Even today we are learning that many simple benzene derivatives are not what they have been thought to be. Many claims stand that will not bear examination. The age of a claim is no index to its respectability: it may be that nobody has found the time to examine it or to venture into a possible criticism of it.

In repeating an experiment from the literature, first consider the place and date of the work. A factory often has better tools at hand for pulverizing, and distilling, for example, than a laboratory. The date of the work may lead to a question as to the purity of the materials employed. Possibly the contaminants to be expected in the materials of today are different, methods of manufacture having changed, from those to be expected at that date. If the quoted work is a patent claim, it must be borne in mind that the claim may be useless or even deliberately misleading. Then consider the question of the purity of your materials. The materials may be better or worse than those described in the literature. When in doubt as to the nature of accompanying impurities in a commercial product, it will be found, as a rule, that the information can be gained directly from the manufacturers. It is now the practice of most reputable manufacturers to supply such information with the understanding that it be not divulged to competitors. In case of failure to repeat the results claimed in a description we should look for some criticism in the literature at a later date. We should consider the matter of impurities, the matter of the terminology in the description, and possibly also the question of catalytic effects of traces of material introduced by some handling of the reaction products as in, shall we say, a washing. Certain points in the description may have to be followed exactly, whereas others may be followed with a liberal allowance of variation: only thought on the matter and a reading on related subjects can help the experimenter.

When using amounts of material that are fractions or multiples of the described amounts, we must not forget to consider a few points that arise. A case of dry distillation will illustrate some of the points. Since the conduction of heat through a mass of organic solid material is slow, and likely to be irregular in various directions, it is almost impossible to assure uniform temperature throughout. Hence a description of a dry distillation which needs to be at all carefully done must involve statements concerning the depth of the layer, the sieving of the particles, the location of the source of heat and a description of it, and a description

Individual Work 119

of the receptacle. It will be realized that a duplication of results with changed quantities may be very difficult even with the best of descriptions. In general, however, a nearly uniform temperature is attained in organic reactions since the reactant mass is usually fluid or partly fluid. Oftentimes the temperature is not mentioned in a procedure, being relatively unimportant. Yet we cannot proceed to change our quantities without considering it, since a reaction in a 1,000 cc. quantity gives 8 times the heat generated in a 125 cc. quantity with a cooling area of surface probably only four times as large. That is, if we had altered the volume of the receptacle in the same ratio as the alteration in mass of material so that an originally half-filled receptacle is still a half-filled receptacle, we would have found ourselves with a cooling area half as effective as in the original. The temperature, in an exothermic reaction, would be consequently higher than in the original experiment, perhaps high enough to produce different results or even to involve danger of explosion.

In changing quantities we should consider the question of time. In general, the time should not be altered merely because quantities are altered, since the relative concentrations (active masses) are unaltered. That is true only on the assumption that the temperature conditions are the same and that the system is a homogeneous fluid system. In a heterogeneous system, however, we can only hold the time constant when we are sure that the surface conditions between the phases are exactly the same as in the original. The usual descriptions leave much to be desired. "Vigorous stirring" may mean that the average particle or droplet is smaller than a millimeter in diameter, or that a certain layer of liquid is extended deeply down the stirring device to the bottom of the flask. It may mean not nearly so much. A decision must be made on such a point. Even when an end point test is given, it must not be assumed that the time factor is unimportant: it may well be that a certain slow side reaction becomes a serious matter when the time is prolonged greatly above the original, even though no mention has been made of such a possibility. To sum up: a change in amounts may or may not involve a change in conditions, but it always involves a question concerning such factors as time, temperature, particle size, and the like.

When initiating new experiments the literature must be surveyed for practical work already done in or near the field of experimentation. This applies not only to new syntheses, preparations of new compounds, and so on, but also applies even to points of technique which are new to the operator. Thus, if a sodium amalgam is being prepared for the first time, it is not sufficient to take the directions of a laboratory manual. The preparation in the manual may entail certain impurities that would be harmful in this particular case, and it is at least probable that a reading concerning sodium amalgam in such practical books as those by Lassar-Cohn and Vanino and Houben-Weyl would lead to a useful understanding of the preparation. It is certain that a failure could not then be blamed on a poor amalgam. Omission of practical detail very often means that the procedure is simple and well known, but it may also mean that the author glossed over some real difficulties in obedience to an editorial policy or a desire for brevity. This is a matter for consideration. Unusual inclusion of practical detail may mean that the detail is very important, even though not explained. It may mean that the detail is one which, when understood, leads to some explanation of some failure or obscurity in your own experiments. Lastly, it should be noted that many failures in the synthetic field are not recorded in the literature. Of late years, many authors have prefaced a description of success with the record of failures. This record of failures is of an importance at least equal to the record of success, from the point of view of most readers on synthetic method. Previously a worker was often obliged to guess at the work preceding a success and to some extent he must still guess. It is still wise to question a promising and plausible suggestion for some synthesis —particularly of a valuable compound—on the likelihood of its being among the unrecorded failures. It is possible that a certain fear of a later success on the same lines by another worker deters many from recording failures: it is possible that many editors feel such records to be a waste of space. Whatever the reason, it is obvious that the organic worker on the synthetic field suffers, while the habit persists, from an impoverishment of facts. The organic literature, as a result, reads like the bulletins of an army. Between the lines the reader must try to find the detail of the event. First read in great detail, noting inclusions and omissions. Be prepared with a knowledge of modern tools for reactions (the synthesis of thyroxin illustrates how useful a modern method of reduction proved to be at a critical analytical point). Lastly, remember the unrecorded failures.

Exercise 19

THE PREPARATION of benzene by dry distillation.—Grind together in a mortar 20 g. of benzoic acid and 40 g. of soda lime. Distill in a small flask with a moving flame. Separate the benzene in the distillate, dry with calcium chloride, decant, and distill the benzene again in contact with a few small pieces of clean sodium. Yield 8 to 9 g. This technique of dry distillation is, in the example above, very crude. It would be

Individual Work 121

better to look around in the literature for examples wherein specific details are given concerning the mesh used for sieving the mixture of dry material, the temperature of the mass, the kind of receptacle, and so on. Such an example will draw attention to the fact that this kind of operation is hard to control or duplicate. The problem of heat conduction through a mass of organic material is the root of the matter, wherefrom arises a necessity for describing the operation in meticulous detail such as depth of material, height of flame, dimensions of receptacles. It will be seen that an exact duplication of conditions is almost impossible when the amount of material differs from the amount used in the original description.

Exercise 20

PHENOL can be obtained by the electrolytic reduction of nitro-benzene (laboratory texts).

Exercise 21

CYCLOHEXANE can be obtained from benzene by reduction with nickel (laboratory texts).

Exercise 22

MANDELIC ACID can be resolved (laboratory texts).

Exercise 23

ORTHO DI-IODO BENZENE can be prepared from ortho iodo aniline which can be obtained from ortho iodo nitrobenzene. Ortho nitraniline can be used as a starting material. See the literature. In the last step of the preparation it is probable that the solid carbon dioxide now available can be used very effectively for two purposes. The yield can probably be made better than the yield described in the literature. Plan this.

Exercise 24

THE USE of bird-shot sodium is involved in Fittig synthesis of ethyl benzene and its homologues (laboratory texts).

Exercise 25

DIETHYL TARTRATE can be prepared by a method which aims at the continuous removal of the water generated in the reaction between the acid and the alcohol. The preparation involves the construction of a simple piece of apparatus (see the literature c. 1900).

Exercise 26

RETENE, or diphenyl, can be prepared from rosin or cyclohexanone respectively by a process involving dehydrogenation with sulphur (see the literature).

Exercise 27

ETHYL GLUCOSIDE can be prepared by a process involving the use of a .25% solution of HCl in alcohol, sealing a bomb tube with alcohol therein, and some careful crystallization (see the literature).

Exercise 28

PHENYL ETHYL BARBITURIC ACID ("luminal" "phenobarbital") is a common hypnotic. It can be prepared from benzyl cyanide and oxalic acid by a process described by Rising and Stieglitz, *J. Am. Chem. Soc.*, **40**:723. There are many points in the procedure which require care. It is best to start with a few hundred grammes of material.

Exercise 29

REVIEW the literature concerning the preparation of b-(p-hydroxyphenyl) ethyl amine, excluding the method from tyrosine and starting from simple cheap materials such as benzyl chloride or cyanide. Check every step with a history of analogous preparations. Note the remarks concerning the value of each synthesis. When any step appears laborious, suspicious by reason of detail, or different from the usual technique in such a method of nitration, reduction, or what not, make careful note of the step. This experiment may fail in the first instance because of lack of experience in evaluating the ease or difficulty of such steps. Try alternative schemes of synthesis. Look for any criticisms of the syntheses up to the present. It is on such matters as these that the student must be alert and somewhat suspicious. The same sort of experience can be gained by trying a scheme of preparation of almost any commercially valuable drug intermediate or drug.

Exercise 30

PREPARE some of the following after noting their uses:
  (a) adrenalone, alanine, arsanilic acid, alloxan, atophane, anthranil, neoarsphenamine (neo-salvarsan), antipyrin, amytal, ascorbic acid
  (b) barbituric acid, benzoflavine, benzylbenzoate, benzidine
  (c) chloracetocatechol, citronellol, cryogenin, carvacrol, a cyanine dye, cephalin, camphor (from pinene), m-cresol purple, chloramine T,

Individual Work

creatinine, cineol, a caffeine derivative, cupferron, coniine, cyclopentenyl acetic acid
- (d) diethoxy acetone, diacetone alcohol, a dipeptide, dibromindigo, distearin, dinitrosalicylic acid
- (e) ephedrine (or homologue)
- (f) formamide, a furane analogue of a benzene derivative
- (g) a glycine anhydride, a thiol aminomethyl glyoxaline, guaiacol, geraniol
- (h) hexyl resorcinol, histamine, hydrastinine, "H" acid
- (i) isoamyl isovalerate, indigotin, indophenol blue, ionone, indigo
- (l) luminal, luminol, lecithin, lysidine
- (m) mercupurin, methyl ionone, malachite green, mandelic acid, an artificial musk, mercurochrome, methylene blue
- (n) novocaine, a novocaine homologue, neutral red, naphthol yellow, yeast nucleic acid
- (o) orthocaine
- (p) piperonal (from isosafrole), phenacetin, pyromucic acid, pyramidon
- (r) a resin, retene, b-resorcylic acid
- (s) stovaine, styrene, salvarsan, saccharine
- (t) tyrosine, thymol, thymol blue, Tyrian purple (a dibromindigo), thiocol, trichlorethyl alcohol (from chloral)
- (u) uracil
- (v) veronal, a veronal homologue, vanillin

Part IV

Use of Literature

A List of Books

THE LIST of books given below is by no means complete. It is drawn up with the intention of helping the student of organic chemistry only in the preliminary stages of reading in his own subject matter and of choosing his private library.

GENERAL

Handbooks

Chemiker-Kalender, Chemical Annual (Nostrand), and *Handbook of Chemistry and Physics* (Chem. Rubber Pub. Co., U. S. A.), provide tables of physical constants, formulae, recipes, etc.

Dictionaries

Thorpe, Ullman, Fehling and Wurtz are authors of dictionaries of general chemical information on the more important inorganic and organic compounds. Heilbron, *Dictionary of Organic Compounds* is limited to the names, formulae, and selected methods of preparation of organic compounds with the exclusion of dyes.

Tables

The most authoritative compilations of physical constants, with references for all individuals, are provided in *International Critical Tables*, *Physikalisch-Chemische Tabellen* (Landolt-Bornstein), and *Tables Annuelles Internationales*. The first is written in English, the second in German, and the last in French. Tables of solubilities of organic compounds only are provided in Seidell's *Solubilities of Organic Compounds*.

ORGANIC CHEMISTRY

Descriptive and General

Beilstein's *Handbuch* is a comprehensive record of all known organic compounds. It is discussed in a separate chapter.

Meyer and Jacobsen's *Lehrbuch* (6 vols.) is a large German text book arranged in the usual order of classes, giving individual and class reactions. Fully referenced, it provides the most complete description of class reactions to be found in a text book proper.

Richter's *Organic Chemistry* (2 vols.; trans. from German)—not to be confused with Richter's *Lexikon*—is a handy descriptive text for advanced students. One volume texts by Karrer (in German), Bernthsen (trans. by Sudborough), Cohen, Conant, Kipping, Lowy and Harrow, and many others are available on the market.

Theoretical and General

Cohen's *Advanced Organic Chemistry* (3 vols.) deals with "Reactions," "Structure," and "Synthesis."

Henrich's *Theories of Organic Chemistry* (trans. from German) has no descriptive material.

Special Theory

On valence problems of organic chemistry Lewis's *Valence and the Structure of Atoms and Molecules*, Sidgwick's *Electronic Theory of Valency*, and Sugden's *The Parachor and Valency* are available.

On structure, Stewart's *Stereochemistry*, Lowry's *Optical Rotatory Power*, and Freudenberg's *Stereochemie* (3 vols.; German) are suggested for collateral reading.

General Practice

Lasser-Cohn and Houben-Weyl have written German texts in methods of organic chemistry, referenced, which deal with methods and apparatus. The chapters describe general operations such as oxidation, reduction, and condensation with references to variations of procedures found useful in particular cases and classes.

Vanino's *Preparative Chemistry* is a collection of preparations in inorganic and organic chemistry that is still sometimes useful.

Meyer's *Constitutions-ermitteilung* describes in detail the reactions and methods most used in the examination of structure of organic compounds. It is incomplete on recent work.

Special Practice

Organic Syntheses is a periodical with cumulative index which gives exact details of convenient or fruitful methods of synthesis of compounds likely to be useful to organic chemists in research work. The selection is left to research workers themselves. Special practice in other branches of chemistry will be found in texts on the several branches.

General Analysis for Groups and Elements

The Identification of Organic Compounds (Mulliken) is a comprehensive scheme for qualitative organic analysis with tabulated tests for some hundreds of common compounds.

Kamm, Clarke, and others are authors of class room texts on the

Use of Literature

same subject. On quantitative analysis, Pregl's *Micro-analysis* describes quantitative methods for small quantities (3 to 5 milligrammes) of material and texts by Thorpe and others describe the quantitative analysis (the ordinary macro-analysis) on 200 milligrammes or thereabouts. Chapters on this subject appear in many laboratory manuals of organic chemistry.

General Analysis

Allen's *Commercial Organic Analysis* and Lunge's *Technical Methods of Chemical Analysis* are comprehensive works (8 to 10 vols.).

Reviews

The periodicals *Annual Reports on the Progress of Chemistry* and *Chemical Reviews* and also the texts *Recent Advances in Organic Chemistry* (Stewart), *Recent Advances in Bio-chemistry* (Pryde), and *Recent Advances in Chemotherapy* (Findlay) should all be available to the organic chemist.

Dyes

Color Index, Farbstofftabellen, Zwischenproducts Teerfabrikation (Lange) are reference books on individual dyes and intermediates. Smaller books on dyes from particular classes of organic compounds are available.

Patents

Friedlander's *Fortschritte Teerfarbenfabrikation* (about 20 vols.) gives a complete list of German synthetic organic patents since 1877 with cumulative index and comments (see below).

Materials

Chemical Guide Book (Chemical Markets, N. Y.) lists the addresses of companies which manufacture a trade material. Reference to the catalogues of the companies gives further information. Since small stocks of special chemicals are sometimes not listed in the catalogues, this book does not provide a complete guide and must be supplemented by inquiry of the individual companies or institutions likely to have a stock of material.

Medicinals

Fortschritte der Heilstoffchemie covers German patents and contains valuable discussions of medicinals. *Constitution and Pharmacological Ac-*

tion (Oswald) and *The Chemistry of Drugs* (Evers) should be mentioned.

BIO-CHEMISTRY

Most research in organic chemistry in modern times is inspired by some problem in medicine or is justified by its importance in life processes generally. For collateral reading, texts should be available on physiological chemistry (Bodansky, Best and Taylor, Bayliss, etc.) ; on anatomy (Gray, etc.) ; on histology (Bailey) and neuro-anatomy; on food chemistry; enzyme chemistry; sugar chemistry (Tollens-Elsner, etc.) ; and on pathology. Texts on colloid chemistry (Freundlich, Kruyt, Thomas, etc.) should also form part of the individual library.

INORGANIC AND PHYSICAL CHEMISTRY

The larger compendiums such as Mellor's *Treatise on Inorganic and Theoretical Chemistry* (15 vols.) and Gmelin-Kraut (25 vols.) which give a description of all known compounds in inorganic chemistry would be needed only for reference.

Representative books for study in physical chemistry include texts by H. Taylor (3 vols.), W. C. M. Lewis (3 vols.), and Eucken (trans. and adapted by Jette and La Mer), and also *Atoms, Molecules and Quanta* by Ruark and Urey and *Thermodynamics* by Lewis and Randall.

THE LIBRARY

A library is usually equipped with a card catalogue of its contents. There are cards for author and subject, and sometimes for the title of a book. Often the catalogue of periodicals is separate from that of books. In a purely chemical library it is likely that much-used sets of lexicons, tables, dictionaries, will be found in some prominent position, together, not in the stacks. The system of shelving may not be indicated on the card and the periodicals may be shelved apart from books: nevertheless the catalogue numbers and shelving will progress systematically as a whole or in each room or division of the library. Hence, if a catalogue number of a book or its subject is known, the book will be easily found. Books on each subject will be arranged alphabetically according to the author's name. A common cataloguing system is the Dewey (decimal) system, wherein Science is 500, Applied Science 600. The classification for Biographies of Chemists is 925.4, that of Bibliographies of Chemistry is 016.54. In the Science division, Mathematics is 510, Physics 530, Chemistry 540, Geology 550, Botany 580 (and Plant Chemistry 581.6), Zoology is 590, Domestic Economy is 640 (food therein is 643), Chem-

Use of Literature

ical Technology is 660 (food therein is 663). In Chemistry itself, General Chemistry is 540; Theoretical and Physical, 541; Handbooks and Tables, 542; Analysis, 543; Qualitative Analysis, 544; Quantitative Analysis, 545; Inorganic Chemistry, 546; Organic Chemistry, 547; Crystallography, 548; and Mineralogy, 549. The subdivisions are usually not important enough to note.

Books Not in the Library

Publishers lists will give sources of the more recent publications. Demands for other books can be made directly to the New York Public Library, to the Library of Congress at Washington, or to other libraries which keep a list of the contents of the national libraries. The Institut Internationale de Bibliographie at Brussels is an international recording office (having no books) of the contents of the libraries of the world.

Periodicals Not in the Library

The Union List of Periodicals (Serials) gives the location of periodicals in the libraries of the United States and Canada. It is supplemented every few years from the reports of the librarians. The Union list is probably at the librarian's desk. The title of the periodical is first sought. It may occur under the name of the society which publishes the periodical. The entry opposite the name gives the location. Thus NNC 1—2(3) —(6)7, 9+. This means that, at Columbia University (NNC being the code for Columbia University shown at the beginning of the List) the periodical is to be found, volume 3 being incomplete, volume 6 also incomplete, volume 8 missing, and the rest all present to date. The list covers dead periodicals and gives a history of the changes of names and the years of existence.

Abbreviated Titles

The usual abbreviations of titles differ from country to country. Some of these abbreviations have been collected: thus the American Chemical Society abstracts for October, 1926, and the Journal of the Society of Chemical Industry for 1924 and 1925 show collections of such abbreviations for the United States and Great Britain respectively. The usual abbreviations used in France and Germany can be found in the journals of these countries (*Bulletin de la Société Chimique de France* and *Berichte der Deutschen Chemischen Gesellschaft*), or, more easily, in the larger descriptive works such as that of Beilstein. Some of the older publications use abbreviations which correspond to a modern abbreviation for a different title: check the index to the abbreviations in case of doubt.

Bibliographies of Chemistry

There are a few books on the market dealing exclusively with the literature of chemistry, and for the less common information concerning dissertations, patents and patent law, the usages of the International Union of Pure and Applied Chemistry, chemical organizations, and so on refer to *Chemical Publications* by Mellon (McGraw-Hill, 1928), *A Guide to the Literature of Chemistry* by Crane and Patterson (Wiley, 1927), or *An Introduction to the Literature of Chemistry* by Mason (Clarendon Press, 1925). Methods for making a search of the literature are described by Hibbert (*Chem. Met. Eng.*, 20:578), and the use of Beilstein is described in *The Use of Beilstein's Handbuch* by Huntress (Wiley, 1930) and (in German) in the foreword to the new Beilstein (Fourth Edition).

References to the Literature

These are given in a conventional form illustrated below. In scientific literature the location of an article in a journal is commonly indicated thus: Jones, R., (title), J. Chem. Soc., [8], 9:356. The article by Jones will then be found in the Journal of the Chemical Society, Series 8, Volume 9, page 356. (In scientific journals the title of the article is usually omitted.)

Book references are given in some such form as: Jones, R., *The Atom* (Macmillan, London, 1916). The title may also appear as "The Atom" or simply Atom.

The accepted form of references to patents is: D. R. P. 45623, Jones & Co., Oct. 6, 1916 (see Friedlander, 8:453). D. R. P. being the abbreviation for Deutsches Reich-patent, this reference is interpreted as one to a German patent issued to Jones & Co. in 1916 and abstracted in Friedlander, Volume 8, page 453.

MAKING A SEARCH OF THE LITERATURE

When the material sought is a question of theory or a subject that may involve excursion into various branches of chemistry, the first requirement is a review of that subject in some book or journal. Reference to the list of books already cited may give a starting point. Reference to publishers' lists and to the card catalogue of a good library may give further information. The material should be collected and arranged on

Use of Literature

cards or in a book in an order appropriate to the subject of the inquiry: it sometimes happens that an author or a time order is a convenient starting method of collection but this should be later brought into an order logical to the inquiry. From the starting material certain references will be gained. If these references are to sources difficult of access (patents of small countries and so on), it may still be possible to find a convenient abstract which will provide all the information needed on that point (see below). The references found will undoubtedly lead to other references, and when in doubt concerning matters of alternative names, conflicting nomenclature, or the place of a title or formula in the indexes of the various books or periodicals, refer to the texts mentioned above, which deal only with the literature. When possible, use the procedure given below for the search for an organic compound. Try alternative titles in perusing any subject index.

When seeking references to a compound of known formula, or to some polymer or product of unknown formula which arises from the compound of known formula, advantage can be taken of the collections of formula indexes. The Beilstein (Fourth Edition) can also be used as described later.

The usual procedure is as follows: If the substance is a salt of an acid or a basic substance, seek it under the mother substance. If it is a simple derivative such as an ester or an oxime, it should be sought under its own formula as well as that of the mother substance. First refer to Richter's *Lexicon* which is a formula index to the known organic compounds up to 1901. The compounds are arranged in order of increasing carbon content and the pages are headed with the number of carbon atoms and the number of associated elements: Thus 9 IV (9 carbon atoms and four other elements), 8 II (8 carbon atoms and two other elements). The order of succession of the elements is C, H, O, N, etc., and all the compounds containing one carbon atom are dealt with before those containing two carbon atoms: compounds containing one carbon atom and three other elements are all disposed of before reaching compounds with one carbon atom and four other elements. Let us suppose we are seeking ortho-nitro cinnamic acid. Written in the order of elements here given, it is $C_9H_7O_4N$. Turning to the pages marked 9 III we proceed by the increase of hydrogen and oxygen atoms to the formula required. We find many other substances of the same formula and the required one under the name b(2-nitrophenyl) akrylsaure Sm 230°C. (meaning melting point 230°C.); Ca + 2 H_2O (meaning that the calcium salt is hydrated with two moles of water); A **163** 192 (meaning that Volume **163** of the *Annalen* has a preparation or other discussion on page 192 concerning this compound; B **13** 2059 (similarly in the Berichte); J 1877, 788 (similarly in a journal whose abbreviation is J, in the year 1877, volume numbers being apparently not used); D.R.P. 21162 (meaning a German patent numbered 21162 and probably abstracted in Friedlander though not so indicated here); M **28** 1164 C 1908 (1) 731 (meaning that the article in Volume 28 of the Journal M has been abstracted in the abstract collection called *Chemisches Centralblatt* whose abbreviation is C, or Cent, or Zentr: if there had been a semi-colon between the two references they would not refer to the same work or article). Finally we see the notation II 1413 II* 854. These final references in Richter are to the Beilstein (3d ed. only; the 4th ed. is not referred to the *Lexicon*). Volume II, page 1413 and the Supplementary Volume II* (Ergangzungsband) page 854 (the pages of the supplementary volumes refer back to the original volume, so it is easy to see which of the two volumes is being handled). On referring to the indicated pages in the Beilstein, further information can be gained about the acid, the patent reference D.R.P. 21162 being followed, without a semi-colon, by the notation Frdl **I,** 29, meaning that the patent has been abstracted or copied into Friedlander's collection of German patents.

From 1910 to 1921, Richter's *Lexikon* is continued in six or seven volumes which must all be consulted: the title of the *Lexikon* in these years may be Stelzner or Stelzner-Richter. The references are, in these volumes, only to the original literature. From 1921 to date, the formula indexes of the *Centralblatt*, the *American Chemical Abstracts*, and the *British Chemical Abstracts* should be consulted. For the current year the individual portions of these abstracts should be consulted. Since these abstracts are sometimes a year or two late the Journals of the various Societies should be consulted, using the subject index where no formula index is available, and finishing with the current portions of the journals. Particular care should be exercised when there is likelihood of the lesser known publications being a source of information on the compound or subject.

Some of the more important Journals (in the abbreviated form recommended by the American Chemical Society) are: Ann.; Ann. Chim. (Ann Chim. Phys.); Ber.; Bull. Soc. Chim.; Compt. Rend.; Gazz. Chim. Ital.; Helv. Chim. Acta; J. Am. Chem. Soc.; J. Chem. Soc.; J. Prakt. Chem.; Montash.; Rec. Trav. Chim. It is perhaps worth noting that the formula indexes in the American publications are not arranged in the usual Richter order. For organic compounds, C comes first, H next, if the substance contains H; with these exceptions the formula is treated as a dictionary of letters where every complexity of letters subsequent is

Use of Literature

listed before the initial letter is changed. The letters are C_1; C_2; C_3; ... H_1; H_2; ... Al_1; Al_2; and the arrangement differs from a dictionary only in that no letter subsequent is a duplicate of a letter precedent.

BEILSTEIN'S HANDBUCH DER ORGANISCHEN CHEMIE
(Fourth Edition)

This edition of Beilstein is likely to be completed before 1940 so that an organic substance of known formula (up to a more recent date than 1921) can be sought in this one book

It is divided into main volumes (banden) dealing with the literature to 1910, with supplementary volumes (ergangzungsbanden) covering the period from 1910 to date: the supplementary volumes may be two or more in number for each original volume. Since the edition is still incomplete the cross-referencing is not done by page and line but by a system number which brings the reader close to the substance sought but not necessarily to the exact page. The system number is found on each page but is of no importance in the plan of the book. The indexes in the volumes are not necessarily a complete record of all the compounds therein and it is best to follow the plan of the book in order to ensure that nothing has been overlooked. This can be done with only a slight knowledge of German. The procedure given here will not be complete but it will serve for most purposes, any doubtful matters being easily solved by reference to the introduction to the Beilstein itself or to some of the books previously mentioned. Desmotropic substances may give a little difficulty. The formula used by the reader, when different from the standard formula used in the Beilstein, is likely to be found in the Beilstein and referred to the system number of the standard form. If not, the introduction to the Beilstein will contain a discussion of it. The standard form used in the Beilstein may be such that certain derivatives must be separated from their related compounds on structural grounds: these matters also are treated in the introduction to the Beilstein, as well as the detail of classification, system numbers, alternative names.

The book is arranged so as to give a definable place to certain simple "index" compounds which can be found either by reference to the index or, better, by a logical analysis of any given compound of known formula. By the method of analysis, indeed, we can find the logical place for a derivative of an index compound even when the corresponding index compound itself is unknown and therefore not tabulated. All compounds of known constitution are considered to be derived in a formal way, as explained below, from these index compounds whose place is known: these derived substances may or may not be found in the index of the volumes.

There are four divisions in the book. The first deals with acyclic (non-cyclic) compounds; the second, with isocyclic compounds (containing rings composed of carbon links only); the third, with heterocyclic compounds (containing rings with one or more links made of atoms other than carbon "hetero" atoms so-called, or "inorganic" atoms); and the fourth, with naturally occurring compounds or materials of unknown formula. The third division is again divided into subdivisions according to the nature and number of the "hetero" atoms involved. In order to decide what division or subdivision contains a given compound, the formula of the compound should be reduced by replacing all elements other than carbon or hydrogen by the equivalent of hydrogen, provided only that the element is directly attached to carbon and that it does not constitute by itself a link of a closed chain. This operation will reduce the formula to one or more of the types indicated. The compound will fall in that division or subdivision of the book which is the latest represented in the root form to which the compound has been reduced.

Thus
$$\begin{array}{c}\text{CH}\!=\!\!\text{C}\cdot\text{OH}\\ \|\quad\quad\|\\ \text{CH}\quad\text{C}\!-\!\text{CH}\!-\!\text{CH}_2\!-\!\text{O}\!-\!\text{CH}_2\!-\!\text{CHCl}\!-\!\text{CH}_2\!-\!\text{CO}\!-\!\text{CH}_3\\ \diagdown\!\diagup\quad\quad|\\ \text{N}\quad\quad\text{MgBr}\\ \text{H}\end{array}$$

reduces to
$$\begin{array}{c}\text{CH}\!=\!\!\text{CH}\\ \|\quad\quad\|\\ \text{CH}\quad\text{CH}\!-\!\text{CH}_2\text{CH}_3 + \text{CH}_3\text{CH}_2\text{CH}_2\text{CH}_2\text{CH}_3\\ \diagdown\!\diagup\\ \text{N}\\ \text{H}\end{array}$$

The first fragment is a heterocycle (Division III) and the second fragment, acyclic (Division I). The original compound will therefore be found in Division III and this principle of latest position will be found always applicable in all further operations in the classification.

Each division or subdivision is divided further into classes. Class I is the "stem nucleus" of the division, or "root" class. Hydrocarbons form the root class in Division I, cyclic hydrocarbons form the root class in Division II, and simple heterocycles in Division III, with the note that in this last Division the "stem nucleus" must contain any "hetero" atom in its lowest valence, and that, if this "hetero" atom is attached to atoms outside the ring, these atoms must be hydrogen. Classes 2-22 are stem nuclei with a characteristic group attached in place of hydrogen to a carbon atom of the stem nucleus. These groups contain an inorganic element

Use of Literature

and one or more hydrogen atoms. These hydrogen atoms are potentially replaceable by other groups and hence the characteristic groups of these classes are often called "functioning" groups since they contain hydrogen capable of further reaction. Such groups, giving rise to numerous derivatives, are considered worthy of characterizing a class, whereas a compound formed by the introduction of such a group as chlorine in the stem nucleus is not important enough to form the basis of a class. Subsequent to class 22 are the unimportant classes denoted as the end of the list below:

1. Stem nuclei.
2. Oxy compounds—C—OH.
3. Oxo compounds—C(OH)—OH (or equivalently C=O)
4. Carboxylic acids—COOH (or equivalently—$C(OH)_3$).
5. Sulphinic acids—SO_2H.
6. Sulphonic acids—SO_3H.
7. Seleninic and selenonic acids.
8. Amines—NH_2.
9. Hydroxylamines—NHOH.
10. Hydrazines—$NHNH_2$.
11. Azo-compounds—N : NH.
12. Hydroxyhydrazines—$N(OH)NH_2$ or—NH.NHOH.
13. Diazonium, diazo, iso-diazo compounds.
14. Azoxy compounds—$N_2(O)H$.
15. Nitramines, isonitramines—$N_2(O_2)H$.
16. Triazanes—N_3H_4.

17, 18, 19. Triazenes, hydroxytriazenes, azoamino oxides.

20, 21, 22. Tetrazanes, tetrazenes, longer N chains.

23-28. Substances with C bound to elements (excepting N) of the 5, 4, 3, 2, 1 and 8 periodic groups in that order, and excepting those alkali metal compounds such as the acetylides which may be regarded as salts. Sodium ethyl belongs here.

The subclasses comprise compounds containing more than one such functioning group and it is logical, on the rule of classification in the latest position, that the class to which such a compound would be referred is the class represented by the latest in order of such functioning groups. The subclasses are again divided into categories on certain natural considerations which are illustrated below for the carboxylic acids.

> Carboxylic acids
>   Dicarboxylic acids
>   Tricarboxylic acids, etc.
>   Hydroxy carboxylic acids
>   With 3 atoms of oxygen
>   With 4 atoms of oxygen
>   With 5, etc., atoms of oxygen

Oxo-carboxylic acids
  With 3, 4, 5 etc., atoms of oxygen
Oxy-oxo-carboxylic acids
  With 4, 5, 6, etc., atoms of oxygen

These categories are not always treated according to increasing oxygen content, but, since the pages of the Beilstein are headed with a description of the subclass and categories being dealt with, the matter need not be elaborated.

Under these subclasses occur certain cells in order of decreasing saturation; the pages are headed with a cell formula, such as $C_nH_{2n-2}O_4$; $C_nH_{2n-4}O_4$; and so on.

In each cell the order of treatment is by increasing molecular weight and it is thus easy to get the empiric formula being described, such as $C_{10}H_{18}O_4$. (Isomers of this molecular formula are dealt with in an order which is not important to note.)

An example is provided in $NH_2CH_2CH_2COOH$. It is in Division I (acyclic compounds). It contains the class groups $-NH_2$ and $-COOH$ directly attached to a carbon of the root form CH_3CH_3 (the stem nucleus). It is placed in the later class (amines). It is in the sub-class oxy-amines. Its formula is $C_3H_7O_2N$ which is written in the cell form $C_nH_{2n+1}O_2N$. The page headings lead to this and under this we find $C_3H_7O_2N$ and under this the desired compound. It is therefore a very simple matter to find a compound in the Beilstein if it happens to be a simple index compound such as that just described. It is a virtue of the Beilstein classification that we can say that all compounds of known constitutional formula, other than index compounds, are derivatives of these index compounds and can be referred to the corresponding index compound after analysis in a logical way as illustrated below. All derivatives of index compounds are described in the Beilstein before the homologue of the index compound.

Examine the formula:

(a) If it shows a valence change from the lowest valence of a characteristic element such as O, N, or S, it can be simplified to the formula with the characteristic element in its lowest valence form. Thus, quaternary ammonium salts or bases, amine oxides, sulphoxides, and sulphones and sulphonium compounds can be simplified to the trivalent N and bivalent S compounds. In the compound Nabcdx it is natural that we simplify to Nabc where a, b, c are the latest groups in class order of the groups a, b, c, d. Certain class groups, such as sulphonic acids, have a characteristic inorganic element in valence higher than its lowest form but class groups are never altered. The present rule must not be applied

Use of Literature

unless the simplification leads to a recognizable class group at that part of the formula or a recognizable functional derivative of a class group. (*b*) If the formula contains more than one functioning group attached to the same carbon atom, or if the formula has a functioning group or groups together with a nonfunctioning group or groups attached to the same carbon atom, it can be simplified in that carbon atom to an —OH group for every such group whether a functional or nonfunctional group. The nonfunctioning groups are —F, —Cl, —Br, —I, —NO, —NO$_2$, —N$_3$, all of which lack hydrogen and so cannot be used for the formation of functional derivatives. They are also, and often, called substituent groups though the term substituent is correctly applied only when these groups are directly attached to C of the stem nucleus. This simplification leads to C(OH)$_2$ or C(OH)$_3$ which are C=O or COOH groups. Thus HC(Cl)(NO$_2$)(NO$_2$) remains HC(Cl)(NO$_2$)(NO$_2$) because there is no functional group on that carbon, but H—C(Cl)(OH)(CH$_3$) becomes H—C=O and H—C(Cl)(OH)(NO$_2$) becomes
          |
         CH$_3$

H—COOH. Likewise H—C(OH)(NH$_2$)(SO$_3$H) becomes H—COOH. One functional group must be present and the sum of functional and nonfunctional groups must be two or more. This rule must not be applied to other than uncomplicated functional and nonfunctional groups. (*c*) When the formula contains nitrogen triply bonded to carbon (—C≡N) it can be changed to —COOH. When the formula contains nitrogen doubly bonded to carbon it can be split into two fragments giving =O to the carbon and H$_2$ to the nitrogen. Thus —C=NC$_6$H$_5$ → —C=O + NH$_2$C$_6$H$_5$. These fragments are separately considered. If one fragment is inorganic it need not be considered. Thus —C(Cl)=NH can be changed to —C(Cl)—OH and NH$_3$ and we then see that —C(Cl)—OH can be operated on according to rule *b*. It becomes —C=O and NH$_3$. The NH$_3$ is neglected. (*d*) Where "substituent" groups occur, the formula is reduced to one with —H in the place of each substituent group. If the carbon to which the "substituent" or "nonfunctioning" group is attached is also bound directly to another inorganic element the "substituent" group is not disturbed. Thus $CH_2Cl\text{—}CHO$ becomes CH_3CHO and $CH_3\underset{\diagdown OH}{\overset{H \diagup}{\underset{}{\overset{\diagdown}{C}}}}\underset{}{\overset{Cl \diagup}{}}$ is operated on according to rule b to become CH_3CHO but $CH_3\text{—}\underset{\diagdown NHC_2H_5}{\overset{Cl \diagup}{\underset{}{\overset{}{CH}}}}$ is left untouched because the same carbon is attached to N and this N is not a part of a simple functional group or simple substituent group. (e) When the formula contains S, Se, Te, it can be reduced in certain circumstances to the same formula with oxygen in the place of S, Se, Te. The S, Se, or Te must be bivalent in the original formula in order that this rule may apply. Thus $-\overset{\overset{S}{\parallel}}{C}-S-C_2H_5$ to

$-\overset{\overset{O}{\parallel}}{C}-O-C_2H_5$ but $CH-\overset{\overset{O}{\parallel}}{S}-O-CH_3$ is untouched. Such compounds are called "replacement" derivatives of the corresponding oxygen compound. (f) The formula may not yet represent a simple index compound. In most cases, however, the formula now represents a simple derivative (a "functional derivative") of an index compound. These derivatives are considered to arise, formally, by "anhydrosynthesis" (loss of water) between H (or H_2) of the functioning group and —OH (or =O) of some organic compound with a carbon attached to —OH (or =O) or by "anhydrosynthesis" between H (or H_2) of the functioning group and some inorganic hydrogen compound. To reduce the formula the formal synthesis must be reversed, splitting between carbon and an inorganic element (usually oxygen) so as to give —OH to carbon and —H to the inorganic element. When the inorganic element is attached to two carbon atoms or more (the usual case) the split must be done on both sides of the inorganic element (not simultaneously since that operation could split beyond an index compound by oversight) and the fragments are all considered.

Thus: $CH_3\text{—}\overset{\overset{O}{\parallel}}{C}\underset{a}{\text{—}}O\underset{b}{\text{—}}C_2H_5$ is split at a or b to give $CH_3\overset{\overset{O}{\parallel}}{C}\text{—}OH + C_2H_5OH$ and $CH_3COOH + C_2H_5OH$. The products are identical in this case but in the case of $CH_3\text{—}\underset{a}{NH}\text{—}\underset{b}{C_2H_5}$ the scission at a leads to $CH_3OH +$ $C_2H_5\text{—}NH_2$ and the scission at b leads to $CH_3NH_2 + C_2H_5OH$. All these fragments come into consideration as possible index compounds. In such a case as $C_6H_5\text{—}CO\underset{a}{\text{—}}O\underset{b}{\text{—}}C_2H_4\underset{c}{\text{—}}O\underset{d}{\text{—}}CH_3$ the scission at a gives $C_6H_5COOH + OHC_2H_4\underset{c}{\text{—}}O\underset{d}{\text{—}}CH_3$ and the second fragment split at c or d

Use of Literature

leads further to $OHC_2H_4OH + CH_3OH$. Splitting at *b* and *c* must also be done until the partitioning reaches index compounds in order to examine the possibility of new fragments appearing. In this present case no new fragments appear (the C—O—C link needs only one scission) but in general the operation must be done.

Classes 1-4 by anhydrosynthesis with organic —OH compounds give the linkage C—O—C which is safely split whenever seen in a formula. The other derivatives of classes 1-4 by anhydrosynthesis with inorganic hydrogen compounds are for the most part simplified in the previous operations from *a* to *d*.

Classes 5-22 often lead by anhydrosynthesis with organic —OH compounds and inorganic hydrogen compounds to derivatives of a type wherein the characteristic element of the class (S or N) is connected to oxygen and thence to carbon. Thus:

$$CH_3—SO_2—O—CH_3 \quad \text{or} \quad CH_3NH—O—CH_3.$$

Such anhydrosynthesis also leads to types such as $CH_3—NH—C_2H_5$ where a characteristic element is attached to two carbons but the element is not oxygen. The operation of splitting can again be safely performed when only carbon atoms are attached to an inorganic element in its lowest valence. Thus: $CH_3—NH—C_2H_5$ would give $CH_3OH + C_2H_5NH_2$ and also $CH_3NH_2 + C_2H_5OH$. The matter of the scission of such types as $CH_3—SO_2—O—CH_3$ and $CH_3—SO_2—NH_2$ (by anhydrosynthesis between class 6 and an inorganic hydrogen compound NH_3) is more complicated. Some compounds such as $(CH_3)_2N—N=O$ (by anhydrosynthesis of $CH_3NH_2 + HNO_2$) even defy a formal analysis in terms of "attachments," and all these derivatives wherein two inorganic elements are bound together are only safely handled after some acquaintance with organic chemistry and after a scrutiny of the class groups and of the possible types arising by anhydrosynthesis. The succeeding splitting processes are therefore merely indications of procedure.

First try a scission between oxygen and carbon. If it leads to a recognizable class group it is probably correct. Thus $CH_3NH—O—CH_3$ leads to $CH_3NH_2 + CH_3—O—OH$ and also $CH_3NHOH + CH_3OH$. In the first scission, however, the coupling group (the one given the —OH in the scission and therefore the one constituting —OH in the synthesis and named the "coupler" to distinguish it from the one contributing hydrogen) is not among the list of coupling groups mentioned below; the second scission leads at once to recognizable class group and coupler. It is probably correct. By a similar operation $CH_3\underset{a}{—}SO_2\underset{b}{—}O—C_2H_5$ is split at *b* into $CH_3SO_2—OH$ and C_2H_5OH which are recognized at once whereas a split at a would lead to $CH_3OH + C_2H_5—O—SO_2H$ and further splitting would give $C_2H_5—OH + HSO_2OH$. The split at a would be suspect because it does not lead to simple compounds which we would expect after the operations $a, b, c, d, e,$ and f up to this point. The rules for the choice of fragments given below would further direct the choice of index compound. In such cases as $CH_3—SO_2—NH_2$, where no oxygen attachment to carbon occurs, the split would first be tried between carbon and the characteristic element. This leads to $CH_3OH + H—SO_2NH_2$. The second is not in the list of couplers but further split makes it $H—SO_2—OH + NH_3$. A further trial of scission between S and N in the original compound gives $CH_3—SO_2—OH + NH_3$. Again the rules for choice of fragments given below would give the correct index compound.

In such a case as $(CH_3)_2N—N=O$ the scission between C and N leads to $2CH_3OH$ and NH_2NO. Scission between N and N leads to $CH_3NHCH_3 + HNO_2$ and this to $CH_3NH_2 + CH_3OH + HNO_2$. Scission between N and O and subsequent splitting would lead to CH_3OH or CH_3NH_2 or $(CH_3)_2N—NH_2 + O(OH)_2$ or $(CH_3)_3N—N(OH)_2$ or some desmotropic form. Rejecting some as unknown it is still true that even rules for choice of fragments would not be as desirable in such cases as a capacity to recognize the probability of its being a nitroso derivative of a secondary amine.

The rule for selection of an index compound can now be stated: "The index compound is that fragment representing the latest class of index compound after the operations $a, b, c, d, e,$ and f have been performed so as to provide only index compounds and inorganic fragments and after the operation f has been performed at all allowable places but rejecting fragments which were given —OH in any operation f which gave two fragments containing carbon."

The last proviso is illustrated in such a formula as $\underset{a}{NH_2—CH_2}$ $\underset{b}{CH_2—NH}—\underset{c}{CH_3}$. No splitting would be allowed at a since the —NH_2 group is a class group. Splitting at b and c gives $NH_2CH_2CH_2NH_2$; $NH_2CH_2CH_2OH$; CH_3OH and CH_3NH_2. The oxy-amine is the latest class or subclass represented, but in its production by scission it was given —OH and the other fragment was CH_3NH_2. It is not an allowable index compound. The next latest class is represented by the diamine $NH_2CH_2CH_2NH_2$ which was given H when produced by the scission at c. It is therefore an admissible index compound; it is the correct index compound because it represents the latest class in the admissible index compounds. Were no such proviso made, the original compound would

Use of Literature

have to be considered as a derivative of both types and would be sought, by the rule of latest position in case of choice, under the oxy-amine. Yet the general plan of the Beilstein allows for the consideration under each index compound of all derivatives with organic compounds having —OH attached to carbon (the organic couplers) which are not later in position than the index compound itself. Under the diamine, by coupling with CH_3OH, it would therefore be expected that NH_2—CH_2CH_2—$NHCH_3$ should appear, but without this proviso it would not appear till NH_2CH_2—CH_2OH was reached.

All operations are necessary in the following case (CH_3) (C_2H_5) (C_3H_7) (Cl) N—CHCl—CH_2—S(O)—C(=NOH)—CH(Cl)NH_2. By operation a $(C_2H_5)(C_3H_7)$N—CHCl—CH_2—S—C(=NOH)—CH(Cl)NH_2. By b $(C_2H_5)(C_3H_7)$N—CHCl—CH_2—S—C(=NOH)—CHO. By c $(C_2H_5)(C_3H_7)$N—CHClCH_2—S—CO—CHO + NH_2OH and since NH_2OH is purely inorganic it is neglected. By d $(C_2H_5)(C_3H_7)$N—CHCl—CH_2—S—CO—CHO cannot be simplified at this stage because the —Cl (the substituent group) is attached to a carbon having another inorganic element attached. By rule e (C_2H_5) (C_3H_7)N—

$$\overset{Cl}{\underset{}{CH}}\overset{a}{-}CH_2-O-CO-CHO. \quad \text{By rule } f \quad \underset{C_3H_7}{\overset{C_2H_5}{\diagdown}}\overset{a}{\underset{b}{\diagup}}N\overset{c}{-}\overset{Cl}{\underset{}{CH}}\overset{d}{-}CH_2\overset{e}{-}O\overset{}{-}CO-CHO$$

scission is performed at a, b, c, d, e, and leads to C_2H_5OH; C_3H_7OH; CH_2ClCH_2OH and this subsequently by operation d to CHO—CH_2OH; $C_2H_5NH_2$; $C_3H_7NH_2$; COOH—CHO. No other organic fragment appears in operation c so that choice of an index compound is limited to the above fragments. The latest fragment is $C_3H_7NH_2$ which by scission at a and c is given H and is therefore not a coupler but an admissible index compound.

By the preceding operations the index compound is located and it is known that the original compound is found somewhere under the index compound as a derivative of some sort. This derivative appears in the Beilstein before the next homolog of the index compound is described. Thus all derivatives of C_6H_5COOH are treated before $C_6H_5CH_2COOH$ appears. The operations a, b, etc. are all procedures for stripping from the original formula the complexities introduced by the formation of derivatives. The derivatives are of three sorts: functional; "substituent" (a derivative formed in substitution of H on the stem nucleus by a substituent group); "replacement" (a derivative formed by replacing O by S, Se, or Te). Though the definition of "formation" is here only formal it is usually the case that these derivatives are actually formed in practice by a reaction of "anhydrosynthesis," or "substitution" with halogens,

144   Use of Literature

etc., or by "replacement," i.e., using phosphorus pentasulphide, etc., on the oxygen compound.

The Order of Derivatives

Because certain index compounds have many derivatives occupying some dozen of pages of the Beilstein, the order of appearance of the derivatives should be mentioned. First come functional derivatives and under each functional derivative any "second degree" or "third degree" or higher degree derivative arising by further anhydrosynthesis or replacement or substitution of the coupler. Thus $C_6H_5COOC_2H_4OCH_3$ is a second degree derivative by anhydrosynthesis between C_6H_5COOH and $OH—C_2H_4—OH$ to give $C_6H_5COOC_2H_4OH$ and a further anhydrosynthesis between this and CH_3OH to give $C_6H_5COOC_2H_4OCH_3$. Such a derivative would appear under $C_6H_5COOC_2H_4OH$. Such higher degree derivatives are thus easily found. The order of treatment of the functional derivatives is very simply remembered: listed first are organic functional derivatives by anhydrosynthesis with organic couplers having —OH attached to carbon in the order of —OH compounds in the Beilstein itself (C—OH, $C(OH)_2$, $C(OH)_3$); second, those with inorganic couplers having H in their formula in the order H_2O_2, oxygen acids, halogen hydrides, nitrogen compounds; and finally, elements of the 5, 4, 3, 2, 1 and 8 groups in that order. If, in such functioning, a compound appears which falls in a later class of derivatives it is treated in the latter place. Thus $C_2H_5OH + HCl$ could be imagined to give $C_2H_5Cl + H_2O$. C_2H_5Cl falls, however, in the substitution derivatives and is there treated.

Substitution derivatives follow functional derivatives. These substitution derivatives are those in which a hydrogen of the stem nucleus of the index compound is replaced by one of the substituent groups. Other types of compounds containing substituents, such as C_6H_5COCl, are considered to arise by anhydrosynthesis of C_6H_5COOH and HCl and fall under the functional derivatives. They are easily distinguished from substitution derivatives by the fact that the carbon on which they are attached is also attached to another inorganic element which is not a part of a simple substituent group. Thus $CH_3—CHCl(NO_2)$ is a substituent derivative of $CH_3—CH_3$ but $CH_3—CHCl(NHOH)$ is obviously some kind of functional derivative. C_2H_5Cl, as stated, is thus a substitution derivative for Beilstein though in actual practice it is usually obtained by anhydrosynthesis. Under the substitution derivative comes any functional derivative of that substituted derivative and any higher degree derivative of that functional derivative. Thus $C_6H_5COOC_2H_5$

Use of Literature 145

comes under CH_3COOH but $C_6H_4ClCOOCH_3$ or $C_6H_4ClCOOC_2H_5$ or $C_6H_4ClCOOC_2H_4OCH_3$ come under $C_6H_4ClCOOH$. Italicized headings in the body of the pages of Beilstein show the division between the various types of derivatives such as functional, substitution or replacement derivatives.

Replacement derivatives come last by replacement of —O— by S, Se, or Te in the index compound. Under such derivative comes any substitution derivative, or functional derivative, or both, of the replacement derivative. Thus any substance which is a derivative of two or more kinds is found in the last category represented in the kinds appearing.

By way of example the substance analysed in detail previously can be considered. $(CH_3)(C_2H_5)(C_3H_7)(Cl)$ N—CHCl—CH_2—S(O)—C(=NOH)—$CHClNH_2$ gave $C_3H_7NH_2$ as the index compound. The C_3H_7— itself is not substituted and neither has the nitrogen (the characteristic group) been replaced by S, Se, or Te. Replacement derivatives, in fact, can only be found among the groups 1-4 since no other group has oxygen in the class group itself and any other change of coupling group makes the derivative merely one of higher degree. It is therefore a functional derivative of high degree of $C_3H_7NH_2$. We imagine it to be formed by coupling with C_2H_5OH and with (Cl)(OH)CH—CH_2—S(O)—C(=NOH)—$CHClNH_2$. The CH_3Cl making the valence change appears directly after such operation. The second coupler splits into (Cl)(OH)CH—CH_2OH so that we can look for the place of coupling with C_2H_5OH and Cl(OH)CH—CH_2OH. The second coupling group is peculiar; it is an imaginary derivative of CH_2OHCHO by coupling with HCl without loss of water. This seems peculiar at first sight but it must be remembered that wherever C=O groups were mentioned they were considered to be equivalent to $C(OH)_2$ and CH_2OH $CH(OH)_2$ + HCl to $CH_2OHCHCl(OH)$ + H_2O is a typical anhydrosynthesis.

In the case of amines with two functional hydrogens it is possible to obtain two first degree derivatives. With —$C(OH)_3$ it would be possible to have three first degree derivatives. The order in this case is determined by the rule of putting in a new coupler, combining this with preceding couplers, and then duplicating the new couplers.

Two Functioning Groups or More

When an index compound contains two functioning groups (e.g., CH_2OH—CH_2—CH_2—COOH) the order of treatment is as follows: when the two functioning groups are the same, the rule is comparable to the rule for a functional group with more than one functioning hydrogen, viz.; the one functioning group does not couple with the next homolog of the coupler or a next-order coupler until the other functioning group has reached an equal order of derivatives. When the functioning groups are different, the rule is first to dispose of all the derivatives of the simpler functioning group (the lower number in the classes) and then to start alteration of the second group, bringing in all derivative changes of the first group before the next homolog of the coupler is considered on the second group.

Heterocycles

The main subdivisions of the heterocyclic division are: rings with O, two O, three O, etc.; rings with one N, two N, three N, etc.; rings with one N and one O, one N and two O, one N and three O, etc.; rings with N and one O, two N and two O, two N and three O, etc.

When the "hetero" atom is bound to hydrogen and also in the ring, the formulae which arise by replacement of the —H on the "hetero" atom by any other group (CH_3, Cl, NH_2, etc.) are considered to arise by coupling of the H with an imaginary OH compound and are treated as functional derivatives of the simplest representative. They precede, therefore, all other derivatives of the simplest representative and fall into place as functional derivatives of the homologs. If any functioning group is present elsewhere in the formula the hetero atom combination to hydrogen is treated as a functional group in respect of the order of the derivatives. The important case of this sort is a ring NH group which would be treated as an amine group as far as order of derivatives were concerned when another functioning group appeared elsewhere in the formula.

Spirocyclic compounds such as $C_6H_4\!\!\begin{array}{c}CH_2\\ \diagup \diagdown \\ CH_2\end{array}\!\!\!\overset{\displaystyle N}{\underset{\displaystyle OH}{|}}\!\!\!\begin{array}{c}CH_2\!\!-\!\!CH_2\\ \diagup \diagdown \\ CH_2\!\!-\!\!CH_2\end{array}$ are considered to be derived from the later cycle $C_6H_4\!\!\begin{array}{c}CH_2\\ \diagup \diagdown \\ CH_2\end{array}\!\!\!NH$ by salt formation through $C_6H_4\!\!\begin{array}{c}CH_2\\ \diagup \diagdown \\ CH_2\end{array}\!\!\!\overset{\displaystyle N}{\underset{\displaystyle OH}{|}}\!\!\!\begin{array}{c}CH_2\!\!-\!\!CH_2\\ \diagup \diagdown \\ CH_2\!\!-\!\!CH_2\end{array}$ Two heterocycles bound together by a carbon chain are considered to belong under the later of the two heterocycles.

Use of Literature

TAUTOMERIC COMPOUNDS

One form is chosen as an index compound and the other form is also found in the Beilstein with a cross reference to the chosen form. Some examples are shown below:

(a) Keto-enol COCH$\rightleftharpoons$C(OH)=C   The index compound is usually the keto form. Exceptions are the aromatic ring, quinoid ring, and pyridine ring compounds, and also any other heterocyclic rings in which the —OH group is not situated on the carbon atom next to the hetero atom. Keto-enol tautomers by ring formation (oxo-cyclo desmotropes)

$$CO{\cdots}\underset{OH}{CH} \rightleftharpoons \underset{OH}{C}\diagup\overset{O}{\diagdown}CH$$

are found under the keto form.

(b) Amino-imino $\underset{NH_2}{CH=CH} \rightleftharpoons \underset{NH}{CH_2-CH}$ These are treated exactly as the keto-enol tautomerism in which the imino form corresponds to the keto form.

(c) C-nitroso-isonitroso $\overset{H}{\diagup}C-NO \rightleftharpoons C=NOH$ The isonitroso (oxime) formula is the index compound except when the nitroso formula is known definitely to be correct. P-nitrosophenols and p-nitrosophenylamines are considered under the oxime formula.

(d) C-nitro-isonitro $C\overset{H}{\underset{NO_2}{\diagup\diagdown}} \rightleftharpoons C=N\overset{O}{\underset{OH}{\diagup\diagdown}}$ are always found under the nitro form.

(e) Nitrosamine-isodiazo   NH — N = O $\rightleftharpoons$ — N = NOH   These occur under the isodiazo formula.

(f) Nitramine-isonitramine   NH—NO$_2$ $\rightleftharpoons$ N = N$\overset{O}{\underset{OH}{\diagup\diagdown}}$ These occur under the isonitramine formula.

(g) Azo-hydrazone   HC—N = N—R $\rightleftharpoons$ C = N—NHR   These occur under the hydrazone formula except when the rearrangement gives an aromatic oxy-azo, or aromatic amino-azo, compound. This preference in the case of aromatic oxy-azo compounds is not to be exercised if the oxy (or amino) form conflicts with the placement of the oxy (or amino) group under keto-enol or amino-imino tautomerism. The usual hydrazone formula is also replaced by the azo-formula in certain cases where rule (*e*) is thus observed. The amino-azo and oxy-azo compounds referred to must be of the aromatic azo-type such as

$$\begin{array}{c} \text{CH}\!\!-\!\!\!-\!\!\!-\!\!\!-\!\!\!-\!\!\text{C}\!-\!\text{OH} \\ \parallel \quad\quad\quad\quad \parallel \\ \text{CH} \quad\quad\quad \text{C}\!-\!\text{N}=\text{N}\!-\!\text{C}_6\text{H}_5 \\ \diagdown\quad\diagup \\ \text{NH} \end{array}$$

and not such types as

$$CH_3\!-\!C(OH)\!=\!C\!-\!N=NC_6H_5 \quad \text{or} \quad \underset{\text{OH}}{\underset{|}{\bigcirc}}\!\!-\!\!N=N\!-\!CO\!-\!NH_2$$

Part V

Qualitative Analysis

Qualitative Analysis

DURING THE waiting periods involved in preparative work on projects taken from the literature, it is advised that substances from the collection of "unknowns" be taken from the stock room and be examined with a view to identification.

The exercise is one of considerable interest in many cases and is normally required of students (pre-medical, pre-engineering, etc.) who will have no opportunity for an extended acquaintance with methods of qualitative analysis in graduate courses.

The discussion given below is intended for those who need only a slight acquaintance with the subject: when difficulties arise, it should be amplified by reference to textbooks on the subject in the library such as those by Kamm, Clarke, Mulliken, Shepherd, and others.

The procedure consists of the observation of the solubility and other physical constants of the material and an examination of its behavior with selected reagents. The characteristic group (or groups) in the compound is thus discovered and its identity with some known substance established by a mixed melting point or better by the preparation of a simple solid derivative of the material having known physical constants.

Some specimens are easily identified after (1) examination of physical constants, (2) solubility tests, and (3) reaction tests; others will need (4) a test for the elements present; all of them are best confirmed by (5) the preparation of a derivative rather than a mixed melting point with a specimen of the suspected substance: therefore, these steps will be described below in that order. The specimens may need (6) purification, but they can be assumed to be pure enough for the purposes of identification unless otherwise noted by the instructor.

PHYSICAL CONSTANTS

Melting Point

This observation is the first one to be made on solids. If the melting point is not sharp, the substance should be purified by crystallization from a suitable solvent until the melting point as taken in the ordinary way extends over less than 1°C.

When taking the melting point it is sometimes useful to continue the heating of the bath to some point about 50°C. above the melting point, or, if the melting point is lower than 100°C., to heat the bath to about 150°C. The behavior of the substance during the heating is carefully observed and recorded.

If the substance does not melt but decomposes at a certain point, that point should be recorded and the observation of this decomposition point repeated in the following way. The bath is heated to a point about 30°C. below the observed decomposition point and the thermometer is then attached to the capillary tube and replaced in the bath. The rate of heating is made fast (about 10°C. a minute rather than the usual 3°C. a minute). Decomposition points are so unreliable, in general, that it is advisable to take a decomposition point with a fast rate of heating and to introduce the capillary tube when the bath is already hot so as to heat the material rapidly. This method will be found to give the most reproducible results.

If the substance first melts and then decomposes a few degrees above the melting point (5°C. to 20°C.), it is to be noted that the melting point may be liable to variation because it is so near the decomposition point. That is to say it will not be as reliable an index to purity as the usual melting point would be.

If the substance vaporizes before it melts it may be purified by sublimation.

If the substance vaporizes at any temperature below 150°C., or decomposes at any point below 150°C., it must not be forgotten when dealing with solvents later on that a loss of substance through vaporization or decomposition may be possible in most hot solvents. Thus a substance decomposing at 150°C. may, near 100°C., decompose to a certain extent by heat effects alone, and thus may appear to be changed by treatment with boiling water. A substance decomposing near 150°C. is likely to be decomposed on steam distillation, for instance, and would, in general, not be treated with the solvents boiling above 80°C. without first testing its stability to them.

If the substance first melts and re-solidifies, and then at higher temperatures melts or decomposes again, we may be dealing with two forms of the same substance or with a change by heat to a higher-melting substance of a different kind. The operator should be prepared to examine both solids for reaction properties, solubilities, etc., later on.

It is sometimes useful to re-melt a substance after having observed its melting point and allowed it to solidify. The melting point should be unchanged. If it is changed, it may mean that: the original melting was complicated by some change of the material; some solvent had been

Qualitative Analysis

driven off during the first melting; or that a rearrangement or decomposition was taking place slowly. If such a phenomenon is observed it would be necessary to continue melting and solidifying until a new steady melting point had been reached or definite indications of general decomposition had been seen.

It is sometimes noted when dealing with different solvents (as in the purification of a substance for analysis) that a melting point changes with change of solvent in a definite way. Thus from one solvent the melting point may change and from another it may remain steady; from one solvent a decomposition point is observed and from another a melting point, and so on. In such cases we must consider the possibilities of reaction with solvent, of compound formation with solvent, and so on.

If any change other than simple melting or decomposition is observed, the matter must be dealt with more fully. Some material should be placed in a small test tube with a leading tube attached to collect any gases by displacement of water. After heating in an oil bath to some convenient temperature in order to initiate the change, any evolved gases are examined as is also the water through which any gaseous products would pass. It may contain halogen acids, volatile solvents, CO_2, etc.

Boiling Point

On distillation of 5-10 cc. in a small distilling flask, a steady boiling point should be observed except for the slight lag of a degree or so while the first cc. passes over and for a slight superheating of a degree or so when the residue is about 1-2 cc. in volume.

If decomposition occurs it should be noted as before and this temperature should not be approached in any further treatment. The distillation should now be tried under reduced pressure and its boiling point recorded if it distills without decomposition.

If the liquid boils below 200°C. it is liable to be lost in evaporation of the ordinary solvents from any solution thereof; care must be taken to avoid this in any later operations. If the substance boils between 200°C. and 300°C. it is not likely to be lost in any great proportion when solvents are evaporated from its solution, but it may be quite volatile in steam if it is immiscible with water; mixtures thereof with water should not be boiled in the open air for any great length of time. A simple test for volatility in steam of any water insoluble liquid or solid is the following: boil with a few cc. of water in a test tube. Turbidity at the top of the test tube, where condensation of water occurs, indicates volatility in steam.

If the boiling point under reduced pressure is not given at the observed point, an estimate can be made roughly of a 1°C. difference for each 1 mm. difference of pressure near 20 mms. A closer approximation at various temperatures is provided in Duhring's Rule, which states that $\frac{t'p_2 - t'p'_1}{tp_2 - tp_1}$ is a constant where t' and t are the boiling points of two substances at the indicated pressures p_1 and p_2. As a reference substance, hexane is suitable for most organic compounds, and water for most of the acids and alcohols (for hexane, see Rechenberg *Zeit. Phys. Chem.* **95**:154: for water see *International Critical Tables* or those by Landolt-Bornstein). Thus $t'p_1$ and $t'p_2$ for hexane is sought from the tables after tp_1 and tp_2 are found for the given substance (tp_2 could well be at atmospheric pressure and tp_1 at the capacity of the water pump). The constant K for the given substance is thus obtained from two observations and leads to the value of t_x, at any other pressure x, by the use of the found value of K in the formula above.

If the liquid is known to be practically pure, the boiling point can be estimated on less than 1 cc. of liquid by two methods.

In the first method, the material is kept gently boiling in a fairly wide test tube with a thermometer hanging just above the liquid and not touching the sides. The refluxing edge of the liquid can usually be seen and the heating should be arranged to keep the refluxing edge steady at a point about half an inch above the top of the bulb of the thermometer.

In the second method, the liquid is placed in a narrow test tube immersed in a bath of liquid in a beaker. A thermometer is kept in the bath which is stirred to provide uniform heating. Inside the test tube is placed a capillary tube which has been heated in a small hot flame at a point about half an inch from the bottom so as to cause the walls of the capillary tube to fuse together at that point but to leave the bottom end unchanged. The bath is brought to a temperature slightly above the boiling point of the liquid for a minute or two and the liquid allowed to boil. The flame is then removed. The bubbles cease to appear at the edge of the capillary tube and the liquid rises in the capillary tube to a point level with the level of the liquid in the test tube. The temperature of the bath at this point is taken to be the boiling point of the liquid in the test tube.

Specific Gravity, Refractive Index, etc.

These are rarely required but a specific notation is sometimes needed as a confirmatory test of sugars or sugar derivatives. Exact solubilities are hardly ever required.

Qualitative Analysis 155

SOLUBILITY TESTS

Exact solubilities are seldom required or used as confirmatory tests for individuals, but rough solubilities, particularly in a group of solvents known as reaction solvents, are very useful in the search for the characteristic group in a compound. Even the general rules of solubility behavior are often useful in the inquiry. These are:

A substance is more soluble in structurally related solvents than in others. The most unlike substances being considered as hydrocarbons and water, the compounds containing a large percentage of carbon by weight would be most soluble in solvents of the hydrocarbon type, while substances containing a large percentage of oxygen as hydroxyl or carbonyl or carboxyl groups would be most soluble in solvents of the water type (the lower alcohols, aldehydes and acids).

The higher the molecular weight, the less the solubility in the usual solvents of low molecular weight. The solubility of resins and other substances of very high molecular weight in certain solvents of low molecular weight (acetone, acetic acid, chloroform) is an exception to this rule. Certain cases of this sort have been considered to be due to loose compound formation or to the formation of colloidal dispersions rather than a true solution.

The behavior with "reaction solvents" is summarized thus:

If the substance is soluble in cold dilute HCl but not soluble in water, it is likely to contain an amino group whose hydrochloride is soluble in water. Insolubility does not mean, however, the lack of an amine group since certain amines do not form hydrochlorides or give very insoluble hydrochlorides. A few other groups lead to solubility in dilute hydrochloric acid. Insolubility is defined for the purposes of this rough work as a solubility of less than one part in twenty or thereabouts.

If the substance is insoluble in water but dissolves in cold dilute sodium or potassium hydroxide it is likely to contain one of the following groups—COOH, SO_2H, SO_3H, phenolic —OH, —SO_2NH_2, —CO—NH—CO—, =NOH, primary or secondary —NO_2, a CO group capable of enolisation, and replacement derivatives by sulphur of the oxygen of certain of the preceding. The alkali salts are formed and dissolve in the water.

If the substance is insoluble in water and in dilute HCl or alkalies but soluble in cold concentrated sulphuric acid, it may be an oxygen compound of the alcohol, aldehyde, ketone, acid or ester type. This solubility is apparently of the reaction type, an "oxonium" compound being formed which is soluble in concentrated sulphuric acid but is very easily hydrolysed by water. In dilute acids the equilibrium is apparently so far disturbed between "oxonium compound" and free oxygen compound that little or no oxonium compound is present and the oxygen compound shows its normal behavior of insolubility in water.

REACTION TESTS

1. Ignite a small portion of substance on a platinum foil, or, if not available, on a crucible lid. Do not use platinum if Pb, Hg, As are present. Carefully note the odor on burning and examine any residue for metallic radicals. A smoky flame suggests aromatic compounds, a luminous nonsmoky flame, aliphatic compounds. A clear nonluminous flame suggests simple aliphatic compounds, a burned-sugar odor, a sugar.

For the purposes of the following tests all substances are divided into categories on the basis of their elementary analysis. If the substance is not found without such an analysis, this must be made later and the tests given below must be referred to the proper category containing (A) C,H,O; (B) C, H, N; (C) C, H, S; (D) C, H, Halogen; (E) C, H, N, and S. It is assumed that all organic compounds contain C and H and these are not tested for in the ordinary course of events.

2. The substance dissolves in cold water to give a strongly acid solution. It suggests

(A) Lower aliphatic acids or a group such as —OH contributing to water solubility in the higher acids, e.g., hydroxy or polyhydroxy acids, phenolic acids.
(B) Some nitrophenols or acid ammonium salts
(C) Sulphonic acids, sulphinic acids
(D) Halogen substituted acids
(E) Nitrosulphonic acids

3. The substance dissolves in cold water to give a faintly acid solution.

(A) Some phenols
(B) Some oximes, some salts of weak organic bases
(C) Sulphonic acids and sulphinic acids
(D) Halogen substituted acids
(E) Nitrosulphonic acids

4. It dissolves in cold water and the solution is neutral.

(A) Lowers alcohols, aldehydes, ketones, glycols, glycerols and sugars
(B) Amino acids, lower amides, ammonium or other nitrogen containing salts of acids

Qualitative Analysis

(C) Some aliphatic sulphoxides
(D) Halogen substituted aldehyde, ketones, alcohols in (A)
(E) Thiocarbamide, salts of nitrosulphonic acids, salts of thiocyanic acids, etc.

5. It dissolves in cold water to give an alkaline solution

(B) Aliphatic amines, some aromatic diamines

6. The substance appears not to dissolve in cold water but dissolves in hot water and the solution is acid, basic, or neutral. The foregoing tests are valid but the substance is probably of higher molecular weight. Sometimes, however, the hot water, or even cold water, only dissolves the substance after decomposition: note any sudden temperature change.

7. The substance dissolves in cold sodium hydroxide but not in water.

(A) Acids, phenols, keto-enol tautomers of certain kinds. Some esters—and also lactones and anhydrides—hydrolyze and dissolve, particularly on warming. Simple acids can be distinguished from phenols by evolution of CO_2 with sodium bicarbonate.
(B) Some amides, amino carboxylic acids, nitro-carboxylic acids, oximes, nitrophenols
(C) Sulphonic acids, sulphinic acids, some mercaptans and those of group (A) combined with sulphur
(D) Halogen substitution products of group (A) and acid chlorides, etc. with decomposition
(E) Nitro-, amino-, sulphonic acids and sulphonamides of the type RSO_2NH_2 and RSO_2NHR

8. The substance does not dissolve in cold sodium hydroxide but does so on warming. If the material comes out unchanged, the observations under the preceding test would generally apply. Some substances undergo decomposition and dissolve (esters, anhydrides, aldehydes). An odor of ammonia suggests ammonium or substituted ammonium salts, thiocarbamides or sulphonamides. Some halogen compounds decompose with water (acid chlorides) or with warm sodium hydroxide (alkyl halides) but others do not decompose even with alcoholic potassium hydroxide. The water solution after warming with alkali should be tested for halogen with nitric acid and silver nitrate.

9. Reaction with concentrated sulphuric acid. On heating with cold concentrated sulphuric acid the oxygen-containing compounds, as stated before, are mostly dissolved. Some substances, however, dissolve in cold

concentrated sulphuric acid only with reaction (olefinic hydrocarbons, some easily sulphonated aromatic hydrocarbons, some esters, some hydroxy acids) and it is therefore necessary to throw the solution on ice for dilution and to establish the recovery of unchanged material when observing its solubility in cold concentrated sulphuric acid.

When heated with concentrated sulphuric acid most substances react or are decomposed except a few aromatic compounds resisting sulphonation and aliphatic hydrocarbons.

10. Reaction with ferric chloride solution. Dissolve the substance in water or alcohol and add one drop of a dilute solution of ferric chloride. Acids usually give a reddish color or a precipitate of the iron salt. Most phenols, keto-enol tautomers give a green blue or violet coloration: oximes, hydroxamic acids also give colors.

11. Dissolve a little of the substance (weighed) in carbon tetrachloride, chloroform, or carbon bisulphide and add drop by drop a solution of known percentage of bromine in one of the same solvents. Immediate decolorization in the cold indicates olefines and the measure of the bromine absorbed before the red color is persistent for a minute or two gives a rough measure of the olefinic unsaturation in the molecule.

Decolorization only after warming, without evolution of HBr, indicates the same unsaturation, but decolorization with evolution of HBr, hot or cold, indicates substances easily substituted such as alcohols, aldehydes, or ketones and phenols. The HBr can usually be detected by breathing across the test tube when a fume is observed from the moisture of the breath.

12. If the bromine test indicates unsaturation, a test with dilute neutral permanganate should be made. Such a solution is decolorized by most olefinic compounds in the cold or with gentle warming.

13. Reflux a portion of the compound with a large excess of a 20 per cent solution of KOH in ordinary alcohol (made by crushing the solid KOH and warming with alcohol) for 20 minutes or more. Dilute with water and distill away the alcohol with a free flame until the distillation thermometer records 100°C. or more. Cool.

This treatment should hydrolyze all esters and decompose to some extent most halogen compounds more reactive than simple aromatic halides (o and p- nitro-aryl halides, acyl halides, alkyl halides). A portion of the water solution can be tested for halogen ion with silver nitrate and nitric acid. The water solution should contain the sodium salt of an acid if esters were present. Addition of sulphuric acid to make just acid should precipitate any insoluble acid. Any soluble acid could be extracted with ether and, after removal of ether, could be examined.

Qualitative Analysis

The distillate of water and alcohol should contain the alcohol group from the hydrolysis of the ester if the boiling point is below or near 100°C. Most esters are methyl or ethyl esters, but by this method ethyl esters could not be detected, as far as the ethyl group is concerned, since ethyl alcohol was used as a solvent. Any odor of ammonia or amines in the distillate suggests an amide group or imide group or an ammonium salt type of linkage in the compound.

14. If the compound is basic to litmus in water or in dilute alcohol solution, or if it shows no solubility in water or alcohol but dissolves in cold dilute or concentrated hydrochloric acid, it is likely to be an amine. Primary, secondary, and tertiary amines can be distinguished by the following test:

Add dilute sodium hydroxide in large excess to a specimen of the material and then a two or three mole excess of benzenesulphochloride (benzene sulphonyl chloride). Warm for a few minutes, with shaking, until all evidence of reaction ceases. Primary amines will remain in solution as the sodium salt of the benzenesulphonyl derivative (R—$NNaSO_2C_6H_5$). Secondary amines will yield insoluble derivatives ($R_1R_2NSO_2C_6H_5$). Tertiary amines will not react. Any insoluble residue is therefore evidence of a secondary or tertiary amine or possibly of a dibenzenesulpho-derivative of a primary amine. The water solution, made acid, should give out an insoluble benzenesulpho-derivative to confirm primary amine. A solid residue must be examined further. The residue is warmed for a few minutes with a mixture of alcohol and potassium hydroxide solution in small proportion. This should decompose any dibenzenesulphonyl derivative of a primary amine to a mono-derivative. Evaporation of the alcohol and dilution with water should not precipitate such mono-derivative (see above) but any tertiary amine should separate from solution, or remain as a residue, and the derivative of a secondary amine should behave similarly. Such precipitate or residue is removed, washed with water by decantation or filtration, and tested with dilute hydrochloric acid. A tertiary amine would dissolve but the derivative of the secondary amine would remain undissolved. If the amine is soluble in water the reaction mixture after treatment with benzenesulphochloride is examined in separate portions (a) for primary or secondary amine as before, and (b) for tertiary amine by extracting well with ether if no primary or secondary amines are present, and (c) for recovery and identification as unchanged material. It is obvious that the method allows even for the identification of mixed amines except that complications are introduced by the presence of water-soluble amines and the fact that the sodium salts of benzenesulphonyl derivatives of primary amines hydrolyze to yield free derivatives in ether extracts.

15. Test for the elements as indicated below. A compound shown to contain combinations of atoms not mentioned previously (e.g., C, H, N, S, Halogen) would be regarded as invalidating some of the tests, but in general it would be considered as merely yielding a sum of the tests. Thus most halogen derivatives would not alter solubilities appreciably since these are usually substituents and it is known that alkyl and aryl halides resemble the hydrocarbons in solubility. On the other hand a C, H, S, Cl combination may be a sulphochloride showing few properties of the usual sulphur containing compounds such as thion or thiol acids, mercaptans, thio esters, sulphonic and sulphinic acids and only yielding convincing evidence after hydrolysis. It is therefore best to hold reaction products aside rather than to dispense with them entirely, until final identification of the material has been made.

TESTS FOR THE ELEMENTS

A small test tube is held in tongs, or fixed in a hole in the center of an asbestos pad supported by a ring, and in it is placed a piece of sodium about the size of a pea. This is fused by heating the test tube and is raised to a dull red heat so that the sodium begins to vaporize. Small specimens of the material are added gradually to the hot sodium and preferably so as to drop directly into the molten sodium. About ¼ g. of material is added in toto and the tube finally raised to a bright red heat. *Spectacles are then donned by the operator.* In order to crack it, the very hot tube is then touched to the surface of some water in a small beaker and is then held just above the water as the mass burns: alternatively the tube is kept at a bright heat for a minute or so and then thrown into the water with hands and eyes out of line of any ejected material. The worst way of performing this operation is to heat to only a low heat and to touch the water gently so as to produce only a small crack in the tube. Heat strongly, do not use too much sodium, bring the tube red hot into the water, protect the eyes.

The contents of the beaker are then ground finely (use a mortar if necessary) and filtered.

Test for nitrogen. A portion of the filtered solution is treated with a few drops of dilute ferrous sulphate solution and then boiled for one minute so as to allow for combination of Fe^{++} with sodium cyanide which should be present if the unknown contained nitrogen. Add one drop of ferric chloride and then acidify. Sodium ferrocyanide combines

Qualitative Analysis

with the Fe^{+++} ion to give blue ferric ferrocyanide $Fe_4[Fe(CN)_6]_3$. The blue color is sometimes slow to appear.

Test for sulphur. To another portion of the filtrate add acetic acid till acid and then a few drops of lead acetate. A black precipitate indicates sulphur as lead sulphide. Alternatively add to this portion of filtrate a drop of a freshly prepared solution of sodium nitroprusside. A violet color indicates sulphur.

Test for halogen. If no nitrogen or sulphur is detected a portion of the filtrate is acidified with nitric acid and a drop of silver nitrate added. If either has been detected, the portion of the filtrate is acidified with nitric acid and boiled for at least five minutes to remove HCN or H_2S from the solution, by decomposition or evaporation. A drop of silver nitrate is then added. A yellowish precipitate is probably iodide or bromide, a white precipitate is probably chloride. A distinction between the halogens is afforded by acidifying a portion of the original filtrate in a test tube in the presence of a little carbon bisulphide, adding drop by drop a solution of chlorine water and shaking frequently. The carbon bisulphide shows a red color if bromine is present, violet if iodine is present, and no change if chlorine is present.

PREPARATION OF A DERIVATIVE

Solid derivatives are to be preferred for ease of purification in small quantities. For alcohols the benzoates or dinitrobenzoates or urethanes are often prepared; for aldehydes, the oximes, semicarbazones or phenylhydrazones; for acids the anilides, amides, or solid esters found from their sodium salts with p-nitrobenzyl bromide or phenacylbromide. For amines, the acetyl or benzoyl derivatives, or picrates, or benzenesulphonyl derivatives are often suitable.

Halogen compounds with reactive halogen can be converted into alcohols (and thence to an ester) or directly into esters with potassium or silver salts of acids. Aromatic halogen compounds must be submitted to some other treatment on other groups attached to the ring.

Esters, amides, acid chlorides, esters, acetals, are usually hydrolyzed and a hydrolytic product converted into a suitable derivative. Nitro compounds can be reduced to amines and the amines identified through a derivative. Sulphonic acids can be converted into sulphonic chlorides and thence to amides.

In the preparation of the derivatives certain conditions must be fulfilled and the best description of the conditions will be found in the literature describing that preparation. A generally useful description of the preparation of a derivative could be given here but it would not apply to a particular case. Thus urethanes are generally prepared by treatment of the alcohol with phenyl isocyanate, avoiding excess, in anhydrous benzene or merely by mixing. The reaction is usually rapid and requires only gentle warming for ten or fifteen minutes even when diluted with a solvent. The urethane is then crystallized from petroleum ether, or alcohol. In particular cases, however, care must be taken to avoid side reactions and difficulty is encountered in purifying from ureas formed by the action of moisture, the isocyanate must be freshly distilled or be taken from sealed stock, etc., etc.

Assuming that the preparation of the derivative has been previously described in full, the duplication of the preparation should give no difficulty if its logic is understood and the properties of by-products are kept in mind. If not described in full, an analogous preparation should be sought in Houben-Weyl or in the Beilstein and conditions arranged in that fashion. It is not safe to assume that, because nitrobenzene is easily reduced by one of a half-dozen methods, the description of a reduction of the nitro compound given in the literature at hand is therefore unnecessarily lengthy or meticulous in calling for certain long heating, or for great excess of hydrochloric acid, or some such item. The assumption may be true, but it is safer to infer that it is not and that this nitro group in the particular compound is difficult to reduce.

Do not look into the open end of test tubes while reactions are proceeding or while heating is taking place.

PURIFICATION

If the specimen is known to be impure, it must be purified to constant melting point as described before or distilled to constant boiling point, and the tests must be performed with certain reservations as to validity. Thus traces of halogens may show up in the halogen test or of nitrogen compounds in the nitrogen test.

If a mixture is to be examined, the separation of the constituents would first be done after a search for a distinction in behavior toward solvents or acids or alkalies. Particularly seek for some distinction in behavior toward aqueous acids or alkalies as these distinctions are usually more clear-cut than ordinary solubility distinctions. Then proceed with the solvents, water, alcohol, benzene, and cold concentrated sulphuric acid as these solvents are usually effective for certain distinct classes of compound. Chloroform, glacial acetic acid, acetone, though often excellent solvents in purification, are not as useful at this juncture.

As stated previously, the solvent chosen for crystallization should be preferably one whose boiling point is below the melting point of the

Qualitative Analysis

solid to be crystallized. This allows the operation of crystallization to be carried out from a solvent which has been refluxed for some time with the solid and has thereby reached its saturation point.

It sometimes happens that a substance comes out from solution as a liquid, later solidifies, and shows a melting point above the boiling point of the solvent. This was previously mentioned in the case of acetanilide and water; the advice to stir well in such cases is based on the nature of the phenomenon. Above 83°C., acetanilide and water in nearly all proportions are in equilibrium in the form of two liquid phases, one of them a solution of acetanilide in water which contains more acetanilide than is soluble at 83°C. and the other a solution of water in acetanilide—the "oil"—which is more dense than the solution. From 83°C. to 100°C. a saturated solution of acetanilide and water can be made containing between 5 per cent and 7 per cent of acetanilide but on cooling this solution gives out above 83°C. only the "oil." As cooling proceeds below 83°C., the solution gives out solid acetanilide but the "oil" loses water to become solid acetanilide. If the "oil" is well stirred at 83°C., it may be possible to achieve a theoretical path for the change from oil to solid via the solution, so that crystallization is perfect. Actually the temperature drops quickly and the "oil" solidifies by loss of water and retains such impurities as are more soluble in the molten material than in the solution. Most impurities would be of such a kind. If, therefore, a saturated solution of acetanilide is made above 83°C. it would be advisable to cool very slowly near 83°C. and to stir vigorously.

Crystallization is usually effected at room temperatures. It is very rarely necessary to cool below 0°C., and thus the freezing point of the solvent itself is very rarely approached (water m.p. 0°C., acetic acid m.p. 15°C., are the only common solvents freezing in this region). When such cooling is necessary near the region of the freezing point of the solvent, however, it must be remembered that certain dilute solutions of a substance in solvent could be cooled to yield solvent rather than substance as the freezing point of the solvent is passed, whereas more concentrated solutions would yield substance in the same region (compare the freezing point diagram for salt in water in inorganic textbooks; see also crystallization, pp. **33, 37, 40, 155**).

Index

Absolute alcohol, 95
Acetamide, preparation, 103
Acetanilide, properties, 33, 36
Acetanilide—water mixtures, 162
Acetanthranilic acid, preparation, 113
Acetone, properties, 42
Acetophenone, properties, 66
Addition tube, 68
Adsorption, 35
Alcohol, absolute, 95
Alcoholic potash, 158
Alcohol—water
  distillations, 71
  mixture, 32
Aluminum chloride, 83
Aluminum sulphate, 54
Amalgams, 80
Amines, test, 159
Aminobenzoic acids, 105
Aniline, preparation, 109
Anisic acid, properties, 33, 36
Apparatus
  requirements, 13, 28
  selection and use, 56, 76
Azeotropic (constant boiling) mixtures, 32, 107

B-(p-hydroxyphenyl) ethylamine, 122
Baths, 28, 29
Beilstein, use of, 135
Benzene, from benzoic acid, 120
Benzoic acid
  from ethyl benzene, 104
  properties, 33, 36
Bleaching powder, 42
Boiling point, 154
  reduced pressure, 78, 153
Brombenzene, preparation, 107
Bromine, 52
  properties, 55
Buchner funnel, 10, 35
Bumping, 13, 74

Camphor, for molecular weight, 106
Capillary tubes, 38
Carbon, decolorizing, 35
Chloroform
  preparation, 41
  properties, 42
Columns, distilling, 88
Condensation, 28, 74
Condensers, 25, 26, 64
Constant-boiling mixtures, 32, 107
Control of reaction, 95
Cooling, 30
Creeping, 48, 88

Crystallization, 33, 37, 162
Cyclohexane, 121

Decolorization, 35
Decomposition point, 152
Derivatives, preparation, 161
Diazonium compounds, 111
Diethyl tartrate, 121
Di-iodo-benzene, 121
Diphenyl, 122
Distillation
  dry, 118, 120
  end point, 29, 83
  fractional, 82, 84, 107
  rate, 22, 31
  simple, 14, 24, 70
  steam, 45
  theory of, 44, 84, 107
  vacuum, 74, 153
Distillation flasks, 24
Distillation of solvent, 72
Drying
  gases, 56, 58
  liquids, 50, 72, 82
  solids, 38, 50
Drying agents, 50
Duhring's rule, 154

Earths, for filtration, 36
Emulsions, 44, 67, 110
End point in reactions, 83, 104
Ether
  distillation, 73
  properties, 67
Ether extractions, 71, 72
Ether-sulphuric acid compound, 30
Ethyl-acetate, preparation, 100
Ethyl-benzene, preparation, 81
Ethyl glucoside, 122
Ethylene, preparation, 51
Ethylene dibromide, preparation, 51
Extraction
  by ether, 71, 72
  continuous, 12
  liquids by liquids, 12, 29, 30
  solids by liquids, 12, 40
  theory of, 13

Ferric chloride test, 158
Filtration, 10, 34, 35, 105
  through cloth, 8
Fractionating columns, 88
Frothing, 43
Funnels
  Buchner, 10, 35
  Hirsch, 10

Index

Funnels—(*Continued*)
  hot water, 34
  separatory, 12
  sintered glass, 10

Gases, washing, 59
Gas generators, 62
Glassworking, 60, 61

Halogens, test, 161
Heating, 13, 28, 29, 118
Hirsch funnel, 10
Hydrocinnamic acid, preparation, 113
Hydrolysis, 158

Impurities, 118
  estimation of, 117
Index compound, selection, 142
Interruption of preparations, 68

Kerosene, 68, 69

Labels, 84
Laboratory note-book, 17, 32
Liquids
  drying, 50, 82
  mutually insoluble, 44
  washing, 63
Literature, use of, 127
Litmus paper, 70
Luminal, 122

Mandelic acid, 121
Manometers, 73, 75
Melting point, 36, 151
  apparatus, 39
  depression, 106
  mixed, 105
Methyl phenyl carbinol, 65
Mixed melting point, 105
Mixing, 23
Mixtures, constant boiling, 32, 107
Molecular weight, estimation, 106
Mutually insoluble liquids, 44

Naphtha, 68, 69
Neutralization, 70
Nitrobenzene, preparation, 108
Nitrobenzoic acid, 104
Nitrogen, test, 160

Olefines, 53, 158
Oxidation, by nitric acid, 104

Patents, 117, 134
Pelegot tube, 57
Phenobarbital, 122
Phenol, 121
Phosphoric acid, 56
Pinacones, 65
Precipitation, hot, 11
Pressing, 36
Procedure, selection, 19, 20, 120

P-tolunitrile, preparation, 111
Purification, general, 162

Quantities, alteration, 82, 108, 118
Quinoline, preparation, 113

Reaction, control of, 95, 119
"Reaction solvents," 155
Reaction temperature, 119
Reaction time, 119
Reaction vessels, 24, 54
Receivers, 27, 74
Reduction, nitro-compounds, 109
Reductions, sodium-alcohol, 68
Reflux ratio, 88
Retene, 122

Safety tubes, 57, 58
Salicylic acid, properties, 33, 36
"Salting out" effect, 101
Sodium, properties, 66, 69, 78
Sodium ethoxide, properties, 66, 67
Solids, drying, 50
Solubility, 101, 162
  tests, 155
Solvents, 33, 40, 155
  distillation, 34, 72
  removal, 102
Steam
  driers, 47
  superheated, 45, 46
Steam distillation, 45
Stirring, 96, 119
Stoppers, 25
Storage, 31
Sublimation, 15, 16
Sulphur, test, 161
Sulphuric acid, solubility in, 157
Sulphuric acid-ether compound, 30, 102
Superheated steam, 45, 46
Superheating, 13

Tea, analysis, 8
Temperature of reaction, 119
Thermometer
  correction, 98
  standardized, 37
Time of reaction, 119

Vacuum distillation, 74, 153
Veronal, preparation, 93
Vessels, reaction, 24, 54

Washing
  gases, 59
  liquids, 63
  on funnel, 10, 36
  solids, 10, 36

Yield
  definition, 21
  improvement, 117

Bei Fragen zur Produktsicherheit wenden Sie sich bitte an:
If you have any questions regarding product safety,
please contact:

Walter de Gruyter GmbH
Genthiner Straße 13
10785 Berlin
productsafety@degruyterbrill.com